ELECTRICAL POWER QUALITY

Other Books by J.B. DIXIT :

1. Mastering C++ Programs (*Published by Firewall Media*)
2. Fundamentals of Computer Programming and Information Technology
3. Programming in C and Numerical Analysis (For B.A. / B.Sc. III)
4. Solutions to Programming in C and Numerical Analysis
5. New Approach to CBSE Computer Science (with C++) XII
6. New Approach to CBSE Computer Science (with C++) XI
7. New Approach to Informatics Practices XI
8. New Approach to Informatics Practices XII (In Press)
9. Excel with Objective Book for NIMCET (MCA Entrance Examination) (*Published by Golden Bells*)
10. Computer Concepts and Programming in C (For U.P.T.U. Lucknow)
11. Fundamentals of Computers and Programming in C
12. Mastering C Programs (*Published by Firewall Media*)
13. Programming in C (*Published by Firewall Media*)
14. Programming in C++ (*Published by Firewall Media*)
15. Programming in C and Numerical Analysis (*Published by Firewall Media*)
16. Computer Fundamentals and Programming in C (For B.Sc. I)
17. Digital Design and Computer Organisation (*Published by University Science Press*)
18. Structured System Analysis and Design
19. Comprehensive Mathematics Activities and Projects (Class IX)
20. Comprehensive Mathematics Activities and Projects (Class X)
21. Information Technology in Business Management
22. *Excel with* Information and Communications Technology
23. Computer Programming and Utilization
24. Fundamentals of Computers
25. Intelligent Instrumentation for Engineers
26. Fundamentals of Computing and Programming
27. Mastering Java Programs
28. Mastering Data Structures Through 'C' Language
29. Presentation Software and Computer Communication
30. Introduction to Computing

ELECTRICAL POWER QUALITY

(For B.E./B.Tech., M.E./M.Tech., and Other Competitive Examinations)

By

J.B. DIXIT
M.Sc. (Applied Mathematics)
Post M.Sc. Diploma in Computer Science,
Kurukshetra University, Kurukshetra,
Haryana

AMIT YADAV
Lecturer, Deptt. of Electrical Engg.
D.C.R.U.S.T., Murthal
Sonepat (Haryana)

UNIVERSITY SCIENCE PRESS
(An Imprint of Laxmi Publications Pvt. Ltd.)
An ISO 9001:2015 Company
BENGALURU • CHENNAI • GUWAHATI • HYDERABAD • JALANDHAR
KOCHI • KOLKATA • LUCKNOW • MUMBAI • RANCHI
NEW DELHI

ELECTRICAL POWER QUALITY

Printed and bound in India
Typeset at : ABRO Enterprises, Delhi
First Edition : 2010, *Reprint* : 2011, 2012, *Edition* : 2020
ISBN : 978-93-80386-74-4

Branches

Bengaluru	080-26 75 69 30	
Chennai	044-24 34 47 26,	24 35 95 07
Guwahati	0361-254 36 69,	251 38 81
Hyderabad	040-27 55 53 83,	27 55 53 93
Jalandhar	0181-222 12 72	
Kochi	0484-237 70 04,	405 13 03
Kolkata	033-22 27 43 84	
Lucknow	0522-220 99 16	
Mumbai	022-24 93 12 61	
Ranchi	0651-220 44 64	

PUBLISHED IN INDIA BY

UNIVERSITY SCIENCE PRESS
(An Imprint of Laxmi Publications Pvt. Ltd.)
An ISO 9001:2015 Company
113, GOLDEN HOUSE, GURUDWARA ROAD, DARYAGANJ,
NEW DELHI-110002, INDIA
Telephone : 91-11-4353 2500, 4353 2501
Fax : 91-11-2325 2572, 4353 2528
www.laxmipublications.com info@laxmipublications.com

C—01193/019/012
Printed at: Ajit Printing Press, Delhi.

Dedicated to

Late Sh. B.R. Yadav

A Source of Inspiration for ever...

PREFACE

We are highly delighted to place the book titled "Electrical Power Quality" in the hands of our esteemed readers.

This book is intended to serve as a textbook for the subject of "Electrical Power Quality" for B.E./B.Tech. degree students. It will also serve as a text reference for the students of M.E./ M.Tech. and Diploma in Engineering. It will also be useful to candidates appearing for AMIE, IETE, GATE, IES and IAS entrance examinations. It would be equally helpful to practising engineers to understand the theoretical aspects of their profession.

Electrical power is perhaps the most essential raw material used by commerce and industry today. It is an unusual commodity because it is required as a continuous flow—it cannot be conveniently stored in quantity—and it cannot be subject to quality assurance checks before it is used. It is, in fact, the epitome of the 'Just in Time' philosophy in which components are delivered to a production line at the point and time of use by a trusted and approved supplier with no requirement for 'goods in' inspection. For 'Just in Time' (JIT) to be successful it is necessary to have good control of the component specification, a high confidence that the supplier can produce and deliver to specification and on time, and a knowledge of the overall product behaviour with 'on limit' components.

The situation with electricity is similar; the reliability of the supply must be known and the resilience of the process to variations must be understood. In reality, of course, electricity is very different from any other product—it is generated far from the point of use, is fed to the grid together with the output of many other generators and arrives at the point of use via several transformers and many kilometres of overhead and possibly underground cabling. Where the industry has been privatized, these network assets will be owned, managed and maintained by a number of different organizations. Assuring the quality of delivered power at the point of use is no easy task—and there is no way that sub-standard electricity can be withdrawn from the supply chain or rejected by the customer.

A perfect power supply would be one that is always available, always within voltage and frequency tolerances, and has a pure noise free sinusoidal wave shape. Just how much deviation from perfection can be tolerated depends on the user's application, the type of equipment installed and the view of his/her requirements.

We have put our sincere effort and knowledge to make you understand the subject-matter in simplest and easiest way. Your valuable suggestions are always most welcome.

—Authors

ACKNOWLEDGEMENT

We express our gratitude to the staff of Deenbandhu Chhotu Ram University of Science and Technology, Murthal (Sonepat, Haryana), for their support in all our efforts towards the writing of this book.

Our heartfelt thanks to Mr. H.S. Chahal, Hon'ble Vice Chancellor and Dr. J.S. Saini, H.O.D. (Department of Electrical Engineering) of D.C.R.U.S.T., Murthal (Sonepat, Haryana) for their constant encouragement in every aspect and providing necessary resources, especially while preparing this book.

We thank our friends at all Universities, Engineering Colleges/Institutes, who encouraged and inspired us directly or indirectly to write a very different and unique kind of book that could serve purpose of all those readers who might be new or an experienced hand in the field of Electrical Power Quality. We would like to mention few special persons who motivated and supported us throughout this process especially Mr. Raman Sharma (Manager—WIPRO), Mr. Naveen Yadav currently working in TCS, Gurgaon, Mr. Deepak Noriyal of Standard Chartered Bank, New Delhi, Mr. Nikhil Popli of Accenture Technologies.

Finally, we wish to express our sincere thanks to our family members and relatives for their moral support.

—Authors

CONTENTS

Chapter 1 INTRODUCTION TO ELECTRICAL POWER QUALITY

1.1. INTRODUCTION

Electrical power is perhaps the most essential raw material used by commerce and industry today. It is an unusual commodity because it is required as a continuous flow—it cannot be conveniently stored in quantity—and it cannot be subject to quality assurance checks before it is used. It is, in fact, the epitome of the 'Just in Time' philosophy in which components are delivered to a production line at the point and time of use by a trusted and approved supplier with no requirement for 'goods in' inspection. For 'Just in Time' (JIT) to be successful it is necessary to have good control of the component specification, a high confidence that the supplier can produce and deliver to specification and on time, and a knowledge of the overall product behaviour with 'on limit' components.

The situation with electricity is similar; the reliability of the supply must be known and the resilience of the process to variations must be understood. In reality, of course, electricity is very different from any other product—it is generated far from the point of use, is fed to the grid together with the output of many other generators and arrives at the point of use via several transformers and many kilometers of overhead and possibly underground cabling Where the industry has been privatized, these network assets will be owned, managed and maintained by a number of different organizations. Assuring the quality of delivered power a the point of use is no easy task—and there is no way that sub-standard electricity can be withdrawn from the supply chain or rejected by the customer.

From the consumers' point of view the problem is even more difficult. There are some limited statistics available on the quality of delivered power, but the acceptable quality leve as perceived by the supplier (and the industry regulator) may be very different from that required, or perhaps desired, by the consumer. The most obvious power defects are complete interruption (which may last from a few seconds to several hours) and voltage dips or sag where the voltage drops to a lower value for a short duration. Naturally, long power interruption are a problem for all users, but many operations are very sensitive to even very short interruptions. Examples of sensitive operation are :

- Continuous process operations, where short interruptions can disrupt the synchronization of the machinery and result in large volumes of semi-processed product. A typical example is the paper making industry where the clean-up operation is long and expensive.
- Multi-stage batch operations, where an interruption during one process can destroy the value of previous operations. An example of this type is the semiconductor industry, where the production of a wafer requires a few dozen processes over several days and the failure of a single process is catastrophic.
- Data processing, where the value of the transaction is high but the cost of processing is low, such as share and foreign exchange dealing. The inability to trade can result in large losses that far exceed the cost of the operation. In a recent example a claim for 15m compensation was made as a result of a 20 minute power interruption.

These are examples of the most sensitive industries, but it is surprising how many apparently mundane operations have quite critical power supply requirements. Examples include large retail units with computerized point of sale and stock control equipment and manufacturing plant with distributed control.

Institute of Electrical and Electronic Engineers (IEEE) Standard IEEE1100 defines power quality as *"the concept of powering and grounding sensitive electronic equipment in a manner suitable for the equipment." As appropriate as this description might seem, the limitation of power quality to "sensitive electronic equipment"* might be subject to disagreement.

A perfect power supply would be one that is always available, always within voltage and frequency tolerances, and has a pure noise free sinusoidal wave shape. Just how much deviation from perfection can be tolerated depends on the user's application, the type of equipment installed and his view of his requirements.

A simpler and perhaps more concise definition might state : *"Power quality is a set of electrical boundaries that allows a piece of equipment to function in its intended manner without significant loss of performance or life expectancy."* This definition embraces two things that we demand from an electrical device : performance and life expectancy.

1.2. POWER QUALITY ISSUES

Power supply quality issues and resulting problems are consequences of the increasing use of solid state switching devices, non-linear and power electronically switched loads, electronic type loads. The advent and wide spread of high power semiconductor switches at utilization, distribution and transmission leaves have non-sinusoidal currents. The electronic type load causes voltage distortions, harmonics and distortion. Power quality problems can cause system equipment malfunction, computer data loss and memory mal function of the sensitive equipment such as computer, Programmable Logic Devices [PLC] controls, and protection and relaying equipments.

Power quality is a two-pronged issue, with electronic equipment playing both villain and victim. Most new electronic equipment, while more efficient than its mechanical predecessors, consumes electricity differently than traditional mechanical appliances. While older devices like incandescent bulbs use power as it is supplied by the utility, electronic devices draw currents in bursts, altering the electricity that flows through them, so that what comes out the other side and returns to the grid is distorted. This "dirty" power underutilizes utility equipment and increases line losses. Thus, some of the efficiency gained through improvements in appliances is lost in the transportation of the electricity that runs them. Additionally, utilities must invest in filters and capacitors to "clean" this dirty power. Poor-quality power also causes transformers, cables and other transmission equipment to burn out more quickly, thus increasing utility equipment costs. In general, the more sophisticated equipment is more sensitive to variations in power quality. Household appliances that were once simple mechanical devices—like furnaces, air conditioners and heat pumps—are going electronic. And the home electronics revolution has made video recorders, personal computers, microwave, ovens and digital clocks (all sensitive to power distortions) common place in homes. As computers get smaller and faster, they become increasingly sensitive to power quality problems.

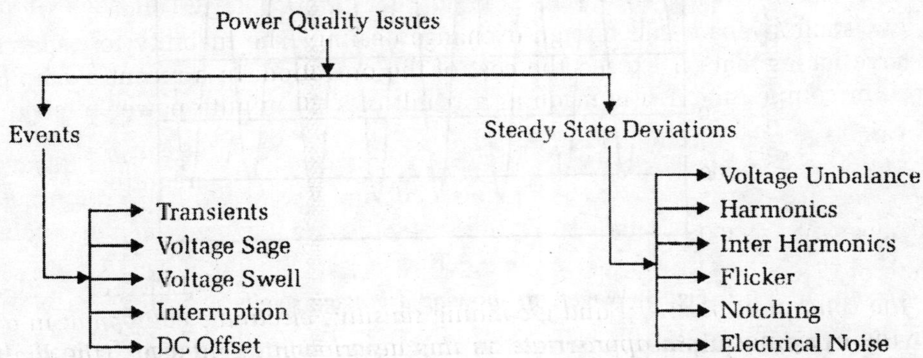

Fig. 1. *Power quality issues.*

1.2.1. Voltage Fluctuations

Voltage fluctuations are changes or swings in the steady-state voltage above or below the designated input range for a piece of equipment. Fluctuations include both sags and swells.

1.2.1.1. Voltage Sag/Dip

By definition a voltage sag (dip) is a reduction in the rms voltage in the range of 0.1 to 0.9 p.u.(retained) for duration greater than half a mains cycle and less than 1 minute. Sag is caused by faults, increased load demand and transitional events such as large motor starting.

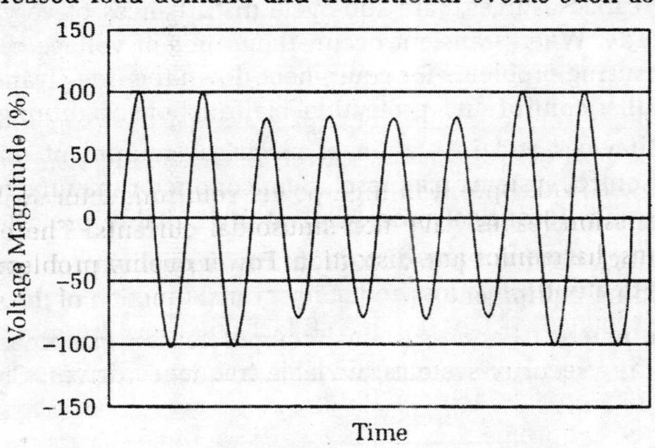

Fig. 2. *Waveform showing a voltage sag.*

1.2.1.2. Voltage Swell

By definition an increase in the rms voltage in the range of 1.1 to 1.8p.u for a duration greater than half a mains cycle and less than 1 minute. It is caused by system faults, load switching and capacitor switching. The magnitude of a voltage swell is dependent upon the fault location, relative to the point of measurement, the system impedance and system earthing. An earthed system with delta-star connected substation transformer will ensure little change in the un-faulted phase's voltages due to a low impedance path from fault to transformer.

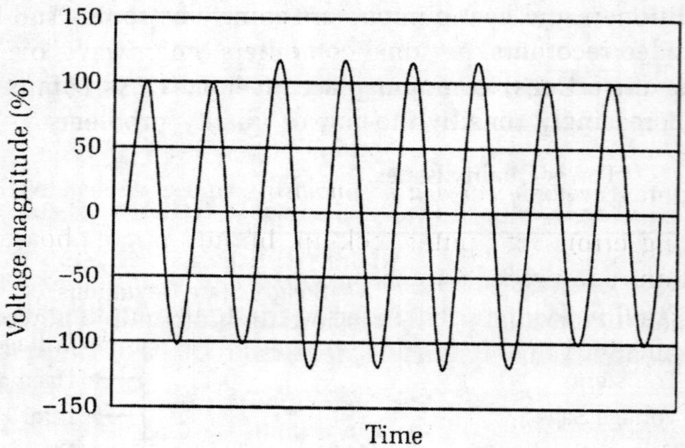

Fig. 3. *Waveform showing a voltage swell.*

- **Causes** : Large equipment start-up or shutdown; sudden change in load; improper wiring; or grounding; utility protection devices.
- **Vulnerable equipment** : Computers; fax machines; variable frequency drives; CNC machines; extruders; motors.
- **Effects** : Data errors; memory loss; equipment shutdown; flickering lights; motors stalling/stopping; reduced motor life.
- **Solutions** : Verify proper electrical connections and wiring; relocate equipment; reduce voltage motor starters; uninterruptible power supply; voltage ride-through equipment.

1.2.2. Transients

Transients, commonly called "surges," are sub-cycle disturbances of very short duration that vary greatly in magnitude. When transient occur, thousands of voltage can be generated into the electrical system, causing problems for equipment down the line. Transients on the supply network are due to faults, control and protection malfunctions, lightning strikes, etc.

Voltage-sensitive devices and insulation of electrical equipment may be damaged, for voltage surges/spikes. Control systems may reset. Semiconductor manufacture can be seriously affected unless the supplies to critical process plant are suitably protected.

- **Causes** : Lighting; normal operation of utility equipment; equipment start-up and shutdown; welding equipment.
- **Vulnerable equipment** : Phone systems; computers; fax machines; digital scales; gas pump controls; fire/security systems; variable frequency drives; CNC machines; PLCs.

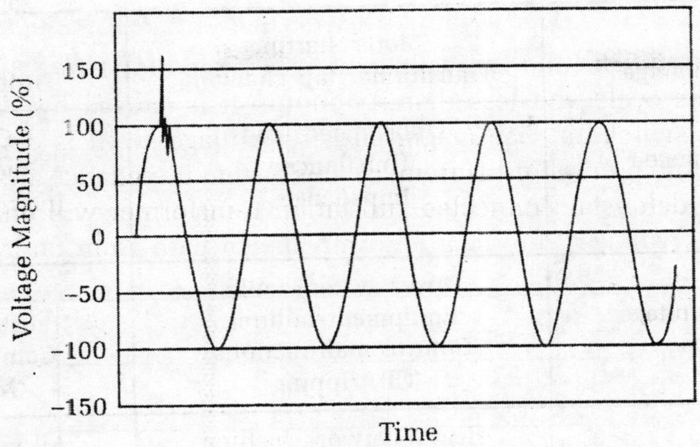

Fig. 4. *Waveform showing an impulsive transient event.*

- **Effects** : Processing errors; computer lock-up; burned circuit boards; degradation of electrical insulation; equipment damage.
- **Solutions** : Transient voltage surge suppression; uninterruptible power supply; isolation transformer; proper grounding.

Table 1 : Power Quality. Issues

Category	*Causes*	*Impacts*
Voltage dips	Local and remote faults Inductive loading Switch on of large loads	Tripping of sensitive equipment Resetting of control systems Motor stalling/tripping
Voltage surges	Capacitor switching Switch off of large loads Phase faults	Tripping of sensitive equipment Damage to insulation and windings Damage to power supplies for electronic equipment
Overvoltage	Load switching Capacitor switching System voltage regulation	Problems with equipment that requies constant steady-state voltage
Harmonics	Industrial furnaces Non-linear loads Transformers/generators Rectifier equipment	Mal-operation of sensitive equipment and relays Capacitors fuse or capacitor failures Telephone interference
Power frequency variation	Loss of generation Extreme loading conditions	Negligible most of time Motors run slower De-tuning of harmonic filters
Voltage fluctuation	A.C. motor drives Inter-harmonic current components Welding and arc furnaces	Flicker in : Fluorescent lamps Incandescent lamps

Rapid voltage change	Motor starting Transformer tap changing	Light flicker Tripping of equipment
Voltage imbalance	Unbalanced loads Unbalanced Impedances	Overheating in motors/generators Interruption of 3-phase operation
Short and long voltage interruptions	Power system faults Equipment failures Control malfunctions CB tripping	Loss of supply to customer equipment Computer shutdowns Motor tripping
Undervoltage	Heavy network loading Loss of generation Poor power factor Lack of var support	All equipment without backup supply facilities
Transients	Lightning Capacitive switching Non-linear switching loads System voltage regulation	Control system resetting Damage to sensitive electronic components Damage to insulation

1.2.3. Interruptions

An interruption is defined as a reduction in the supply voltage or load current to a level less than 0.1p.u. for a time not more than 1 minute. Interruptions can be caused by system faults, system equipment failures or control and protection malfunctions because the magnitude of the supply voltage drops below 0.1p.u. during an interruption, the magnitude of the interruption is considered to be zero, the interruption is defined only by duration. The length of short interruptions is generally governed by the type of protection equipment utilized on the system such as auto-reclosers.

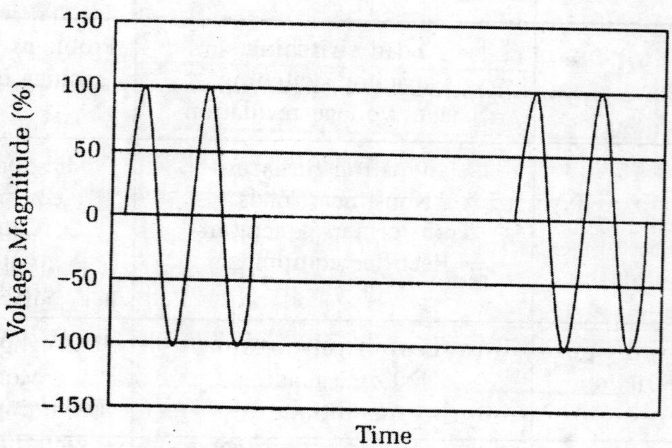

Fig. 5. *Waveform showing a short interruption.*

Longer interruptions—power cuts—are usually thought of as being caused by the supplier but can also is caused by the failure of on-site equipment, conductors and connections. Careful design using high resilience techniques can minimise the effects. The objective is to identify single points of failure and eliminate them by providing redundant equipment or alternative supply paths so that operation can continue despite a single failure. Systems designed in this way are easier to maintain and are better maintained as a result. It is important that maintenance procedures are developed at an early stage as part of the resilient design concept. Standby generation and UPS systems, required to cover short and longer term power cuts are essential elements of a resilient system.

1.2.3.1. Power Outages

Power outages are total interruptions of electrical supply. Utilities have installed protection equipment that briefly interrupts power to allow time for a disturbance to dissipate. For example, if lightning strikes a power line, a large voltage is instantly induced into the lines. The protection equipment momentarily interrupts power, allowing time for the surge to dissipate.

- **Causes** : Ice storms; lightning; wind; utility equipment failure.
- **Vulnerable equipment** : All electrical equipment.
- **Effects** : Complete disruption of operation.
- **Solutions** : Transient voltage surge suppression; uninterruptible power supply.

1.2.4. Electrical Noise

Electrical noise is high-frequency interference caused by a number of factors, including arc welding or the operation of some electric motors.

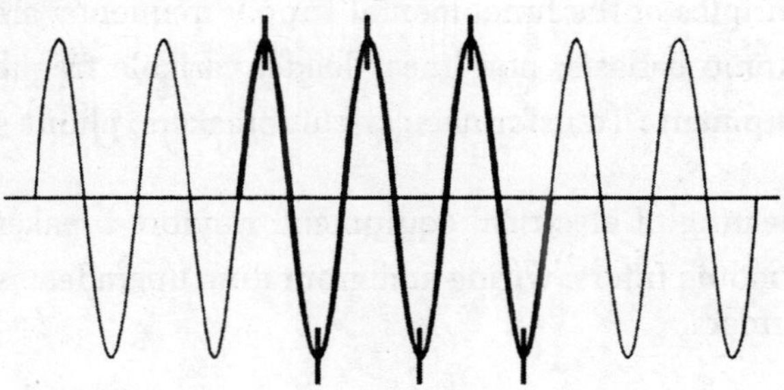

Fig. 6. *Waveform containing electrical noise.*

- **Causes** : Lighting; normal operation of utility equipment; equipment start-up and shutdown; welding equipment.
- **Vulnerable equipment** : Phone systems; computers; fax machines; digital scales; gas pump controls; fire/security systems; variable frequency drives; CNC machines; PLCs.
- **Effects** : Processing errors; computer lock-up; burned circuit boards; degradation of electrical insulation; equipment damage.
- **Solutions** : Transient voltage surge suppression; uninterruptible power supply; isolation transformer; proper grounding.

1.2.5. Harmonics

Harmonics are the periodic steady-state distortions of the sine wave due to equipment generating a frequency other than the standard 50 cycles per second. Harmonics are periodic sinusoidal distortions of the supply voltage or load current caused by non-linear loads. Harmonics are measured in integer multiples of the fundamental supply frequency. Using Fourier series analysis the individual frequency components of the distorted waveform can be described in terms of the harmonic order. magnitude and phase of each component. The quantity Total Harmonic Distortion (THD) is commonly used, it represents a summation of all the harmonic components present in a waveform (n = 1).

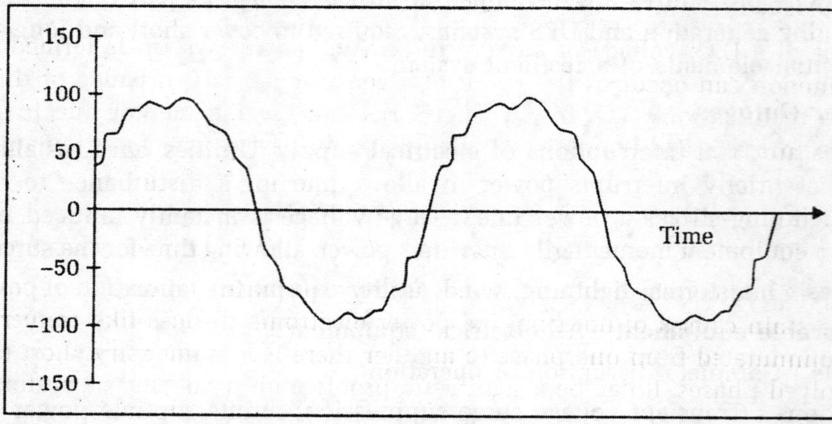

Fig. 7. *Harmonic distortion.*

1.2.5.1. *Inter-harmonics*

Distorted voltage or current waveforms containing periodic distortions of a sinusoidal nature that are not integer multiples of the fundamental supply frequency are termed inter-harmonic.

- **Causes** : Electronic ballasts; non-linear loads; variable frequency drives.
- **Vulnerable equipment** : Transformers; circuit breakers; phone systems; capacitor banks; motors.
- **Effects** : Overheating of electrical equipment; random breakers tripping; hot neutrals.
- **Solutions** : Harmonic filters; wiring and grounding upgrades; isolating non-linear loads; special transformers.

1.2.6. Flicker

A term used to describe the visual effect of small voltage variations on electrical lighting equipment (particularly tungsten-filament lamps). The frequency range of disturbances affecting lighting appliances, which are detectable by the human eye, is in the range of 1 to 30 Hz. Flicker severity is the intensity off licker annoyance defined by the UIE-IEC flicker measuring method and evaluated by the following quantities :

- short-term severity (P_{st}) measured over a period often minutes.
- long-term severity (P_{lt}) calculated from a sequence of 12 P_{st} values over a two hour interval, according to the following expression :

$$P_{lt} = \sqrt{\sum_{i=1}^{12} \frac{P_{st_i}^3}{12}}$$

1.2.7. Voltage Imbalance

Voltage imbalance is defined as a deviation in the magnitude and/or phase of one or more of the phases, of a three phase supply, with respect to the magnitude of the other phases and the normal phase angle (120). Voltage imbalance is more commonly a problem in rural areas where one, or two, phase(s) of a three-phase supply is/are being loaded.

1,2.8. DC Offset

The presence of a D.C. voltage or current in an A.C. power system is termed as D.C. offset. This phenomenon can occur as the result of a geo-magnetic disturbance or due to the effect of half-wave rectification. D.C. in A.C. networks can be detrimental due to an increase in transformer saturation, additional stressing of insulation, waveform distortion and other adverse effects.

1.2.9. Notching

Notching is a periodic voltage disturbance caused by the normal operation of power electronics devices. The main causes of notching are power electronic devices like converters, when the current is commutated from one phase to another there is a momentary short circuit between the two involved phases. It can be a significant problem on weak electric system, where it can produce noise currents causing control system misoperation. It can damage capacitive components connected in shunt due to high rate of voltage rise at the notches.

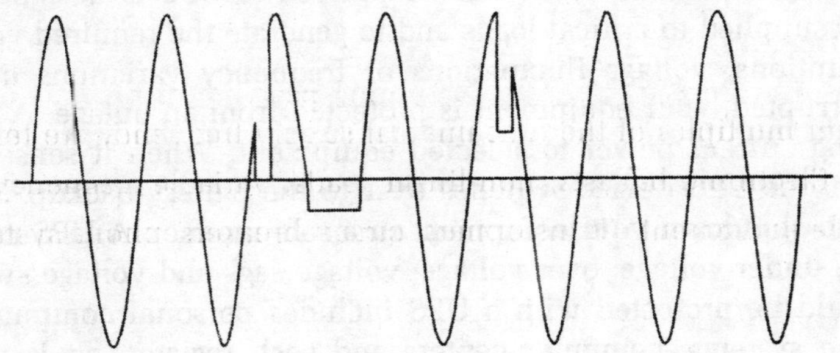

Fig. 8. *Notching.*

1.2.10. Short Circuit

A short circuit (or "short") is not normally considered a quality problem as much as it is a dangerous operational malfunction or fault. Short circuit refers to a condition where two "hot" lines are connected directly (or through small impedance) or one "hot" line is connected directly to ground.

A short circuit causes very high fault currents to flow through the wiring and all devices between the point of the short and the incoming power line. Left unchecked, a short circuit can very quickly lead to catastrophic overheating, melting and burning of wiring and devices. The opening of a breaker or the operation of a protective fuse is the normal means of guarding against damage from shorts circuits. It is imperative that protective breakers and fuses be of the proper size and characteristics to avoid the dangers of short circuits.

1.3. REMEDIAL MEASURES

There are many methods available for correcting Power Quality problems. The most common are given in Table 2. Brief details of each method are given below, but it is emphasized that the solution adopted will be tailored specifically to the problem and site.

Table 2 : Common Methods Available for Correcting Power Quality Problems

Equipment	*Application*
UPS	Voltage variations Supply interruptions Frequency variations
Earthing practices	Harmonics
Filters (Active/Passive)	Harmonics
Energy storage devices	Voltage variations Supply interruptions

1.3.1. Uninterruptible Power Supply (UPS)

This is one of the most common types of protective equipment. It is placed between the electrical outlet and the equipment you wish to protect. A UPS is designed to improve the quality of the power supplied to critical loads and to generate the required voltage and current when power interruptions, voltage fluctuations or frequency variations occur. Even if the utility power is interrupted, your equipment is protected from an outage. A UPS has batteries to provide alternating current power to selected equipment. When it senses a power outage or change in voltage, the UPS device switches over to the battery backup. The batteries give you operating time to shutdown your equipment in a safe manner. A UPS can protect against power interruptions, under voltage, over voltage, voltage sags and voltage swells. Some of the equipment that should be protected with a UPS includes personal computers, workstations, fileservers, telephone systems, computer centers and cash registers. It is very important to know the cause of the problem before deciding if a UPS is required and choosing a UPS that will provide the necessary protection.

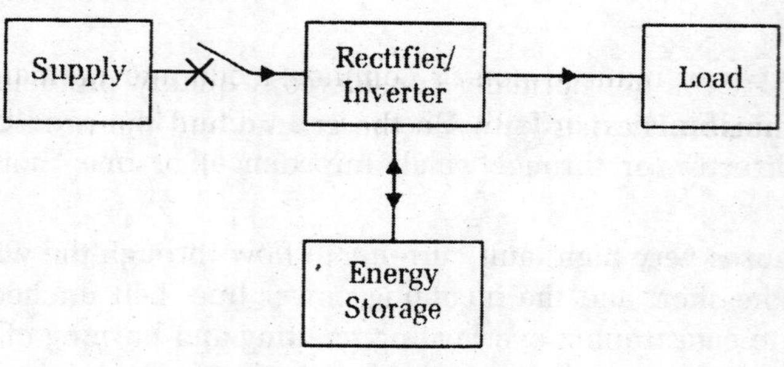

Fig. 9. *UPS system.*

For small single phase applications, the Uninterruptible Power Supply (UPS) is frequently the power conditioner of choice. The UPS can provide power conditioning as well as battery backup in the event of a power interruption. This permits time for ride through of the interruption or for the safe backup or shutdown of systems. The UPS has to be sized for the peak current required by the downstream devices and then for the length of time that full load current is required for system operation or shutdown. The physical size and cost of the UPS grow quickly as the size of the load and "battery time" increase. In large three phase applications, UPS can grow to room size units requiring sophisticated maintenance and monitoring systems to prolong the life of the batteries. Battery replacement in a UPS can be required in as few as 2 years, can run upto 65% of the initial cost of the unit, and depending on the battery type, can require special disposal of the old batteries.

UPS come in four basic versions : the standby (or off line), the line interactive, the ferro-resonant and the double conversion. The standby UPS only offers protection in the event of long sag or an interruption. In its normal mode, the battery and inverter of the standby UPS are off, and the downstream devices operate off of unconditioned line power. When a sag or interruption occurs, a mechanical switch turns on the UPS so that it can provide power to downstream devices. Once the event is cleared, the UPS goes back into its standby charge mode. Many standby UPS also incorporate surge protection. These types of units are most frequently used to protect individual devices like workstations.

The line interactive works the same way as the standby UPS but adds the capability to provide some degree of voltage regulation (buck-boost) of the line power for the downstream devices. The ferro-resonant UPS replaces the buck-boost portion of the line interactive UPS with a ferro-resonant transformer to provide a higher degree of power conditioning of the line power for the downstream devices.

The double conversion UPS provides the highest degree of power conditioning of any UPS. While the previous three types of UPS only utilize the batteries and inverter in the event of a sag or interruption, the double conversion UPS uses two inverters. This UPS takes incoming A.C. power, converts it to D.C. and then converts it back to A.C. power. In the event of a sag or interruption, the output inverter uses D.C. battery power to supply the load. The double conversion, with the backup provided by the batteries, can eliminate address many power quality problems, however this capability comes with a high first cost, operating cost and maintenance cost (**For Detailed Description See Chapter 2**).

1.3.2. Noise Filters

A noise filter eliminates interference that may interrupt the proper operation of electrical or electronic equipment. This is known as Electromagnetic Interference [EMI]. Electronic devices can be a source of this interference.

There are two types of reduced noise : common mode and normal mode.

Common mode noise : It exists between the ground and the neutral. Electronic devices are most sensitive to common mode noise. A shielded isolation transformer is very effective in reducing common mode noise.

Normal (or transverse) mode noise : It exists between the "hot" lines and the neutral. Normal mode noise is generally also reduced with a shielded isolation transformer.

1.3.3. Isolation Transformers

These devices electrically separate the electronic equipment from the incoming power system, reducing unwanted electrical noise. A transformer wound such that each winding is electrically isolated from the other windings, *e.g.*, current will not pass from one winding to the other winding by conduction. This isolation serves to reduce minor transients and noise in the input from being transmitted to the output.

Isolation transformers are often fitted with shielding between the primary and secondary to further suppress noise and transients. These units are primarily used to step up or step down voltage and offer some protection from noise but little else in the way of power conditioning.

1.3.4. Harmonic Filters

A harmonic filter limits the periodic or continuous distortions of the normal voltage wave pattern [sine wave] caused by non-linear loads.

Non-linear loads, including computers and electronic lighting ballasts, draw electric current differently than linear loads and distort the normal voltage pattern. An example of a linear load would be an incandescent light bulb.

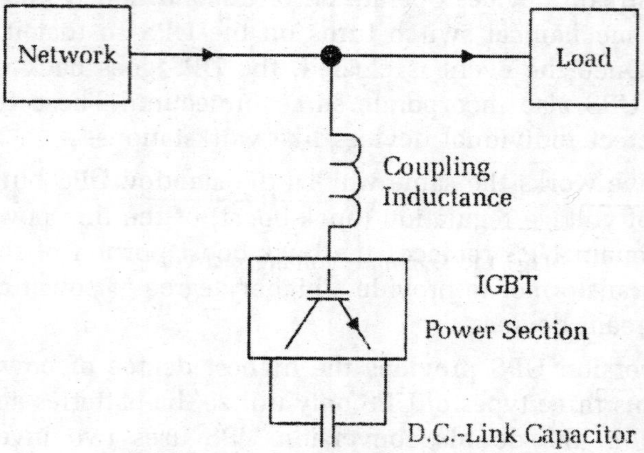

Fig. 10. *Active harmonic filter.*

Harmonic filtering is removal of harmonics from incoming power. Passive filters are designed (tuned) to remove harmonics of specific frequency and amplitude but will provide little protection if the harmonic's characteristics change. Active filters (compensators) analyze the incoming power for deviations from the desired waveform. When a harmonic is detected, the unit superimposes a signal upon the waveform that is the opposite of the harmonic which then cancels out the harmonic. Passive filters are relatively inexpensive but have limited use. Active filters are expensive and are used in special circumstances where harmonics have become a problem. In most cases, harmonics are reduced to tolerable levels through isolation transformers or other power conditioning devices.

1.3.5. Dynamic Voltage Restorer

A Dynamic voltage restorer is a power electronic controller that can protect sensitive loads from disturbances in the power supply. It is a power electronic converter based Device that has been design to protect critical loads from all supply side disturbances other than outages it is connected in series with a distribution feeder and is capable of generating or absorbing real and reactive power at its A.C. terminals.

The basic principle is by inserting a voltage of required magnitude and frequency the DVR can restore the load side voltage to desired amplitude and frequency. Even when the source voltage is unbalanced and distorted usually a DVR is connected to protect sensitive loads during faults in the power supply system. A DVR that can regulate voltage at the load terminals against any variation in the supply side voltage while consuming no real power.

This is a voltage source converter and energy source connected in series (either directly or via an injection transformer) that controls the voltage downstream directly by injection of suitable voltage in series with the source. Ratings of upto several MW are possible at voltages upto 11 kV.

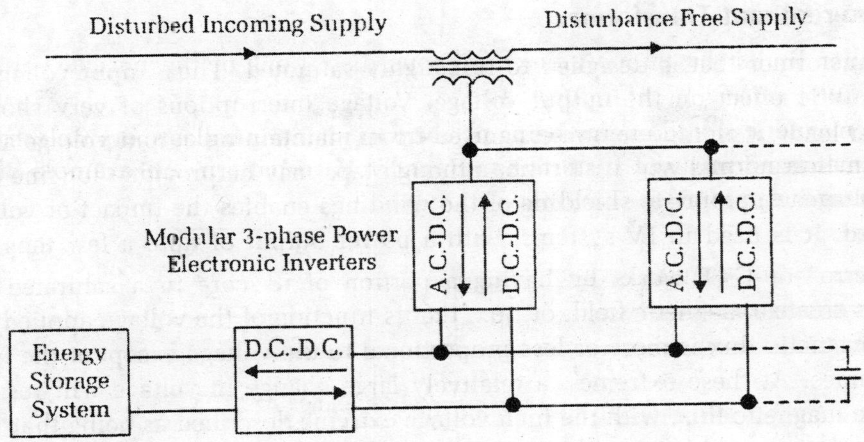

Fig. 11. *Dynamic voltage restorer.*

1.3.6. DYSC

A DYSC is a power electronic device containing a series sag corrector and a shunt converter that provides voltage sag immunity with a minimum retained of voltage of 50%, which covers 92% of the voltage sags that have been reported in a large study.

1.3.7. Flywheel

A flywheel with out a diesel generator protects the equipment against all sags as long as the inertia of the flywheel can support the load. Most flywheels can supply the rated load for 3–15 s, which is sufficient to protect against all voltage sags but not against supply outages.

1.3.8. Static Var Compensator (SVC)

This is a shunt-connected assembly of capacitors, and possibly reactors, which provides reactive power to a network during disturbances to minimise them. It is normally applied to transmission networks to counter voltage dips/surges during faults and enhance power transmission capacity on long transmission circuits. The devices are switched either in discrete steps or made continuously variable through the use of PED's. It works by providing reactive power (leading/lagging as required) to assist in keeping the voltage at the point of connection constant. Voltage variations at that point are reflected in var variations, so provision of reactive power of appropriate sign can reduce the voltage fluctuations. The STATCOM is a SVC comprised of a self commutated static converter and capacitor energy storage. The switching of the converter is controlled to supply reactive power of appropriate sign to the network.

1.3.9. Earthing Practices

A site that suffers from problems with harmonics may need to investigate the earthing of equipment. The high neutral currents that result can give rise to overheating/failure of neutral/earth connections, while high neutral-earth impedances can give rise to common mode voltage problems. All neutral and earth connections need to be checked to ensure they are adequately sized and have sound joints.

1.3.10. Ferro-resonant Transformer

This is a transformer that is designed to run highly saturated. Thus, input voltage dips and surges have little effect on the output voltage. Voltage interruptions of very short duratio result in the magnetic stored energy being used up in maintaining output voltage and current The transformer is normally of 1 : 1 ratio, although taps may be provided for fine adjustment of output voltage. Appropriate shielding of the windings enables the impact of voltage spikes to be reduced. It is used in LV systems, with a power output of upto a few tens of kVA.

The "ferro" or CVT works by having a portion of its core in a saturated condition. Transformers create a magnetic field, or flux, that is function of the voltage applied. To a large extent, the magnetic flux is more or less proportional to the voltage except at the high or low voltage extremes. At these extremes, a relatively large change in voltage will result in little change in the magnetic flux, with the high voltage extreme described as being the "saturation region. An analogy would be that of having a very large tank of water with a small outlet pipe at the bottom. As the water flow into the tank changes, the flow of water out of the small pipe remains fairly constant due to the large volume of water in the tank.

The majority of ferros are used in applications below 1.5 kVA, although there are models upto 25 kVA. Ferros can provide good noise attenuation and are very reliable. On the downside, ferros can add significant harmonic distortion, have a serious problem with high inrush currents (>170%), have very poor efficiency at less than full load and tend to be very hot and noisy in operation.

1.3.11. Surge Suppressors

These are small plug-in devices designed to protect equipment from moderate surges and spikes. Surge suppressors should be considered the minimum level of protection from internal and external transients on electrical, telephone and data lines. Installing surge suppression at the main service panel and following through to each electronic device is strongly recommended.

Surge arrestors are generally Metal Oxide Varistors (MOV) that clip (truncate) voltages above a predetermined threshold. Essentially a non-linear resistor, the surge arrestor is placed between a phase and ground. At normal voltage levels, the resistance of the surge arrestor is very high such that it negligible current flows to ground. At voltage levels above the threshold, the resistance of the surge arrestor becomes so low as to effectively create a short circuit, thereby diverting the damaging power of the surge to ground. Surge arrestors provide valuable protection against dangerous voltage levels.

1.3.12. Voltage Regulation

One of the primary functions of voltage regulators and power conditioners is to maintain the voltage level seen by downstream devices within a certain set of limits. Typically, units will specify "input regulation" or "input range" and "output regulation". For example, one unit might have an input range of +10%/–25% and a regulation to ±5%. This means that if the incoming voltage is within the nominal voltage plus 10% and nominal voltage minus 25%, then the unit will regulate the voltage output to within the nominal voltage plus or minus 5%. If the nominal voltage is 480 volts, then as long as the input voltage is no higher than 528 volts and no lower than 360 volts, then the output voltage will be between 504 and 456 volts. Both input range and regulation will vary by manufacturer, technology and price range. Large input ranges and small regulation ranges tend to have higher price tags and vice versa. Most applications can be satisfied regulation of ±3 to 5%. Mechanical tap changing voltage regulators frequently offer output regulation in the range of ±1/2 to 1%, however these types of units have a significantly slower response time than other technologies.

1.3.13. Guidelines for Power Quality

For getting good power quality the following guidelines has to be followed :

- Install no more than four to six out lets per branch circuit in those branch. Circuits that serve computers and other devices.
- Use over sized conductors to account for peak loads, since those loads may be uptothree times that of r.m.s current levels
- Specify oversized neutrals in 3 phase circuits feeding nonlinear load branch circuits.
- Power *Q* Grounding resistance not greater than 10 ohms. For sensitive equipment systems not to more than 2 ohms.
- Specify copper-wound harmonic-concealing isolation transformers between nonlinear load.

1.4. POWER QUALITY Vs. EQUIPMENT IMMUNITY

All devices are susceptible to power quality; no devices are 100% immune. All electrical power system installations have power quality anomalies to some degree and no power systems exist for which power quality problems are nonexistent. The challenge, therefore, is to create a balance. In Fig. 12, the balanced beam represents the electrical power system.

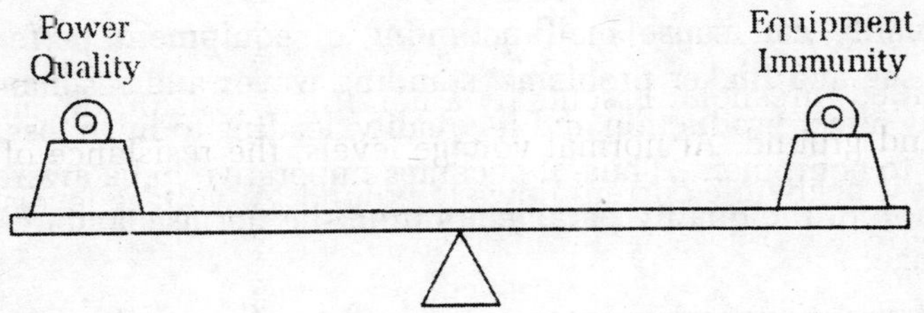

Fig. 12. *Power quality and equipment immunity.*

Power quality and equipment immunity are two forces working in opposition. The object is then to a create a balance between the two. We can assign power quality indices to the various locations in the power system and immunity indices to the loads. By matching the immunity index of a piece of equipment with the power quality index, we can arrive at a balance where all equipment in the power system can coexist and function adequately. Experience indicates that three categories would sufficiently represent power quality and equipment immunity (see Table 3). During the design stages of a facility, many problems can be avoided if sufficient care is exercised to balance the immunity characteristics of equipment with the power quality environment.

Table 3 : Immunity and Power Quality Indices

Index	*Description*	*Examples*
	Equipment Immunity Indices	
I	High immunity	Motors, transformers, incandescent lighting, heating loads, electromechanical relays
II	Moderate immunity	Electronic ballasts, solid-state relays, programmable logic controllers, adjustable speed drives
III	Low immunity	Signal, communication and data processing equipment; electronic medical equipment
	Power Quality Indices	
I	Low power quality problems	Service entrace switchboard, lighting power distribution panel
II	Moderate power quality problems	HVAC power panels
III	High power quality problems	Panels supplying adjustable speed drives, elevators, large motors

1.5. POWER QUALITY CONCERNS

Power Quality and Reliability is the key to successful delivery of quality product and operation of an industry. It is now even more critical to the industry because of increasing application of electronic loads and electronic controllers which are sensitive to the quality of power supplied. These can have serious economic consequences and cost business millions of rupees each year in revenues loss, process improvements, and scrapped product. There is a dire need for all concerned to discuss the business of power quality and the latest technologies for improving power system efficiency and reliability.

Bad power quality can cause malfunctioning of equipment performance. Harmonics, voltage unbalance, sag and flicker problems, standing waves and resonance—are some of the issues that adversely affect production and its quality, leading to huge loss in terms of product, energy and damage to equipment. Thus, it becomes imperative to be aware of quality of power grid and the deviation of the quality parameters from the norms/standard to avoid breakdown or equipment damage.

Automation of the industrial processes has lead to effective increase in production and reduction in cost. Somehow, the controls in automation are so sensitive to power supply that even trivial variation can create huge negative impact on production and product quality. This can be sorted out from the source itself, like transformer, switchgear, relays, is necessary. Latest innovations for askance distribution system design; including—high reliability distribution systems using high-speed fault detection, rapid fault clearing, fault isolation and automatic restoration systems—is not a distant dream. The focus on the dynamic characteristics of the system equipments while facing the abnormal supply should not be lost.

Most of the electrical equipment is an industrial facility requires high-quality electricity, showing zero tolerance towards power outages, no matter how short-lived it may be. Research shows that 80% of all power of all power quality and reliability problems occurs inside end-user's facilities. A timely diagnosis of the current status of equipment can help prevent major breakdown, process interruption and reduce monetary loss. Thus, measuring critical parameters pertaining to the equipment with the right quality instruments is necessary.

Benefits of industrial automation in saving labour cost and enhancing overall efficiency and product and quality are well-known. Companies in a wide range of business are expected to increasingly implement automation equipment. However, the increasing level of automation increases the use of electronic or solid state device, further enhancing the need for the safeguarding the power quality issues. These, if denied, result in the system unreliability and loss in efficiency of the equipment. But this cannot be the reason for not implementing automation, what is required is a proper system while going for automation to take care of the pollutants in the power system. In today's advanced technology, automation and power conservation along with better power quality is not far from reality contributing to savings in energy, money and minimizing human intervention.

Acute power shortage of course has created an alarming situation in the country. This apart, power quality problems like voltage fluctuation, frequency variation, generation of spikes, impulses, surges and sags, harmonics, high earth leakage current, missing cycles, black-outs, brown-outs, etc. are playing havoc on our assets. Besides downtime and production losses they cause incalculable damage to our costly capital equipment.

Power line disturbances such as voltage sag/swell, outages, flicker, harmonics, and incur a heavy loss to electric utility customer due to the wide application of the non-linear and electronically switched devices *n* distribution systems. At the same time, modern industrial equipment are more sensitive to these power quality problems than before and need higher quality of electrical power. The basic reason is that the minor power disruptions which one would have noticed only as momentary flickering in the lights, may now interrupt the whole automated factory assemble lines with sensitive electronic controllers or make all computers in an office go blank at once.

The generally encountered power quality problems are voltage dips and fluctuation, momentary interruptions, harmonics, transients, etc. Expect the power outages and transients, etc. Except the power outages and transients due to lightning, other power quality problems like sags, swells short term interruptions, harmonics, distortions, etc. occur either due to some faults in distribution feeders or interference from loads. During short circuits or fault clearing process in feeder, the neighbouring feeders can get affected with sags or swells in the supply voltages.

1.5.1. The Good Quality of Power

The good quality of power can be specified as :

- The supply voltage should be with in guaranteed tolerance of declared value. In India the specifications related to power quality are ± 10% variation in voltage and ± 2% of frequency.
- The wave should be a pure sine wave within allowable limits for distortion.
- Voltage should be balanced in all 3 phases.
- Supply should be reliable.
- The earthing system should serve its purpose properly.

1.5.2. Causes and Consequences

The causes and consequences of power quality problems can be traced to a specific type of electrical disturbance. By analyzing the waveform of the disturbance, power quality engineers can determine what problems your facility has and what the optimal solution is.

For comparison purposes, a normal voltage waveform is 50 cycles per second—at most plus or minus ten percent of nominal voltage.

Power quality is an increasingly important issue for all businesses. Problems with powering and grounding can cause data and processing errors that affect production and service quality.

- **Lost production :** Each time production is interrupted, your business loses the margin on the product that is not manufactured and sold.
- **Damaged product :** Interruptions can damage a partially complete product, cause the items to be rerun or scrapped.
- **Maintenance :** Reacting to a voltage disruption can involve restoring production, diagnosing and correcting the problem, clean up and repair, disposing of damaged products and, in some cases, environment costs.
- **Hidden costs :** If the impact of voltage sag is a control error, a product defect may be discovered after customer delivery. The costs of losing repeat sales, product recalls and negative public relations can be significant and hard to quantify.

1.6. POWER QUALITY STANDARDS

Power quality is a worldwide issue, and keeping related standards current is a never-ending task. It typically takes years to push changes through the process. One of the most important developments in the power quality arena is the increased emphasis on coordinating IEEE standards with international standards developed by the International Electro-technical Committee.

Most of the ongoing work by the IEEE in harmonic standards development has shifted to modifying Standard 519. So let's take a closer look at some of the recent developments within these two organizations.

The Power Quality Standards Coordinating Committee, SCC-22, sponsored a task force to pull together a list of power quality terms and definitions. However, as the task force began compiling the definitions from various IEEE and IEC standards, they found many confusing or conflicting terms. Despite this hurdle, they tried to identify official definitions and provide examples of properly used terms. Accurate comparisons of power quality levels from one facility and system to another require consistent methodology. Existing IEEE Standard 1159 provides only general guidelines and definitions, so its group is actively developing more specific procedures for systems monitoring. Standard 1159.3 is actually based on the development of a Power Quality Data Interchange Format (PQDIF). The format is a means for exchanging power quality monitoring information between different applications. It will allow software developers to design applications that analyze power quality problems independently from the manufacturers of the monitoring equipment. As for the IEC, there are some specific standards related to the monitoring requirements for each type of power quality phenomena. For example, IEC Standard 61000-4-7 deals with the requirements for monitoring and measuring harmonics, while IEC Standard 61000-4-15 describes the instrumentation and procedures for monitoring flicker. IEC Standard 61000-4-30 have some future plans on providing overall recommendations for monitoring all types of power quality phenomena while still referring to other specific standards where appropriate. IEEE is currently adopting this standardized approach as well.

1.6.1. IEEE 519

IEEE 519-1992, Recommended Practices and Requirements for Harmonic Control in Electric Power Systems, established limits on harmonic currents and voltages at the Point of Common Coupling (PCC), or point of metering.

The limits of IEEE 519 are intended to :

1. Assure that the electric utility can deliver relatively clean power to all of its customers;
2. Assure that the electric utility can protect its electrical equipment from overheating, loss of life from excessive harmonic currents, and excessive voltage stress due to excessive harmonic voltage. Each point from IEEE 519 lists the limits for harmonic distortion at the Point of Common Coupling (PCC) or metering point with the utility. The voltage distortion limits are 3% for individual harmonics and 5% THD.

The heaviest electrical users have the most stringent harmonic limits. However, as the size of the user load decreases with respect to the size of the system, the user can inject a larger percentage of harmonic current into the utility system. This protects other users on the same feeder as well as the utility, which is required to furnish a certain quality of power to its customers. Note that all of the harmonic limits in IEEE 519 are based on a customer load mix and location on the power system. The limits are not applied to particular equipment, although, with a high amount of nonlinear loads, it is likely that some harmonic suppression may be necessary. As the load mix changes it is often necessary to re-evaluate the limits of IEEE 519.

1.6.1.1. IEEE-519 Standard for Current Harmonics

- **General Distribution Systems [220 V–69 kV]**

Below current distortion limits are for odd harmonics. Even harmonics are limited to 25% of the odd harmonic limits. For all power generation equipment, distortion limits are those with $I_{SC}/I_L < 20$. I_{SC} is the maximum short circuit current at the point of coupling "*PCC*". I_L is the maximum fundamental frequency 15-or 30-minutes load current at *PCC*. *TDD* is the Total Demand Distortion (=*THD* normalized by I_L).

1.6.2. IEC 61000-3-2 and IEC 61000-3-4

1.6.2.1. IEC 61000-3-2

It specifies limits for harmonic current emissions applicable to electrical and electronic equipment having an input current upto and including 16 A per phase, and intended to be connected to public low-voltage distribution systems. The tests according to this standard are type tests.

1.6.2.2. IEC/TS 61000–3–4

It specifies to electrical and electronic equipment with a rated input current exceeding 16 A per phase and intended to be connected to public low-voltage A.C. distribution systems of the following types :

- nominal voltage upto 240 V, single-phase, two or three wires;
- nominal voltage upto 600 V, three-phase, three or four wires;
- nominal frequency 50 Hz or 60 Hz

These recommendations specify the information required to enable a supply authority to assess equipment regarding harmonic disturbance and to decide whether or not the equipment is acceptable for connection with regard to the harmonic distortion aspect. The European standards, IEC 61000-3-2 and 61000-3-4, placing current harmonic limits on equipment, are designed to protect the small consumer's equipment. The former is restricted to 16 A; the latter extends the range above 16 A.

1.6.3. IEEE Standard 141-Recommended Practice for Electric Power Distribution for Industrial Plants

A thorough analysis of basic electrical-system considerations is presented. Guidance is provided in design, construction, and continuity of an overall system to achieve safety of life and preservation of property; reliability; simplicity of operation; voltage regulation in the utilization of equipment within the tolerance limits under all load conditions; care and maintenance; and flexibility to permit development and expansion.

1.6.4. IEEE Standard 142-Recommended Practice for Grounding of Industrial and Commercial Power Systems

This standard presents a thorough investigation of the problems of grounding and the methods for solving these problems. There is a separate chapter for grounding sensitive equipment.

1.6.5. IEEE Standard 241-Recommended Practice for Electric Power Systems in Commercial Buildings

This standard deals with the installing electric power system in commercial buildings.

1.6.6. IEEE Standard 242-Recommended Practice for Protection and Coordination of Industrial and Commercial Power Systems

This standard deals with the proper selection, application, and coordination of the components which constitute system protection for industrial plants and commercial buildings.

1.6.7. IEEE Standard 446-Recommended Practice for Emergency and Standby Power Systems for Industrial and Commercial Applications

This standard is recommended engineering practices for the selection and application of emergency and standby power systems. It provides facility designers, operators and owners with guidelines for assuring uninterrupted power, virtually free of frequency excursions and voltage dips, surges, and transients.

1.6.8. IEEE Standard 493-Recommended Practice for Design of Reliable Industrial and Commercial Power Systems

The fundamentals of reliability analysis as it applies to the planning and design of industrial and commercial electric power distribution systems are presented. Included are basic concepts of reliability analysis by probability methods, fundamentals of power system reliability evaluation, economic evaluation of reliability, cost of power outage data, equipment reliability data, and examples of reliability analysis. Emergency and standby power, electrical preventive maintenance, and evaluating and improving reliability of the existing plant are also addressed.

1.6.9. IEEE Standard 1100-Recommended Practice for Powering and Grounding Sensitive Electronic Equipment

Recommended design, installation, and maintenance practices for electrical power and grounding (including both power-related and signal-related noise control) of sensitive electronic processing equipment used in commercial and industrial applications.

1.6.10. IEEE Standard 1159-Recommended Practice for Monitoring Electric Power Quality

Monitoring of electric power quality of A.C. power systems, definitions of power quality terminology, impact of poor power quality on utility and customer equipment, and the measurement of electromagnetic phenomena are covered.

1.6.11. IEEE Standard 1250-Guide for Service to Equipment Sensitive to Momentary Voltage Disturbances

Computers, computer-like products, and equipment using solid state power conversion have created entirely new areas of power quality considerations. There is an increasing awareness that much of this new user equipment is not designed to withstand the surges, faults, and reclosing duty present on typical distributions systems. Momentary voltage disturbances occurring in A.C. power distribution and utilization systems, their potential effects on this new, sensitive, user equipment, and guidance toward mitigation of these effects are described. Harmonic distortion limits are also discussed.

1.6.12. IEEE Standard 1346-Recommended Practice for Evaluating Electric Power System Compatibility with Electronic Process Equipment

A standard methodology for the technical and financial analysis of voltage sag compatibility between process equipment and electric power systems is recommended. The methodology presented is intended to be used as a planning tool to quantify the voltage sag environment and process sensitivity. It shows how technical and financial alternatives can be evaluated. Performance limits for utility systems, power distribution systems, or electronic process equipment are not included.

1.6.13. SEMI F-47-Standard for Definition and Measurement of Equipment Reliability, Availability, and Maintainability

This standard defines the sag ride through capability.

1.6.14. SEMI F-42-1999, Test Method for Semiconductor Processing Equipment Voltage Sag Immunity

This standard defines the test methodology to confirm compliance to the standard.

1.6.15. Standards related to Voltage Sag and Reliability

Personnel at many utilities monitor voltage sags and calculate indices for system sag performance the same way they do for system reliability. The distribution voltage quality standard *i.e.*, IEEE

Standard P1564 gives the recommended indices and procedures for characterizing voltage sag performance and comparing performance across different systems. Undoubtedly, this standardized approach will be extremely useful. Individual customers can use the voltage sag performance information to evaluate the economics of various alternatives for improving system performance.

A new IEC Standard 61000-2-8 titled "Environment —Voltage Dips and Short Interruptions" has come recently. Its purpose is to describe the expectations and characterizations of system performance. This standards warrants considerable discussion within the IEEE to avoid conflicting methods of characterizing system performance in different parts of the world.

1.6.16. Standards Related to Harmonics

Harmonic standards development by the IEEE has shifted to modifying Standard 519-1992. Discussion at the IEEE winter power meeting focused on appropriate limits for harmonic levels inside customer facilities. IEEE Standard 519-1992 provides recommended limits for harmonic levels at the Point of Common Coupling (PCC) between the customer and the power system (*i.e.*, the location where other customers could be supplied). The recommended voltage distortion limit for the PCC is 5% for the Total Harmonic Distortion (THD) and 3% for individual harmonics.

The revision to Standard 519 is considering higher limits for inside the facility and making these limits frequency-dependent. The limits specified in IEC for low-voltage systems allow a THD of 8% and include limits for individual harmonic components, which decrease with frequency. "The Guide for Applying Harmonic Limits on Power Systems," is recently in process to be added up in IEEE Standard 519A. Other activities in the harmonics area is to add up the "Working on single-phase harmonic limits and inter-harmonics issues" in the existing standard. The harmonic filter up, which is part of the capacitor subcommittee, has recently completed a harmonic filter design guide known as IEEE Standard P1531. A new task force on active filters has been formed under the sponsorship of the IEEE harmonics group.

1.6.17. Standards Related to Flicker

Developments in voltage flicker standards demonstrate how the industry can successfully coordinate IEEE and IEC activities. IEC Standard 61000-4-15 defines the measurement procedure and monitor requirements for characterizing flicker. Recent updates to this standard include the response characteristics of incandescent lights on 120 V, 60 Hz systems, which permit using the same monitoring equipment to characterize flicker on power systems in North America. The IEEE flicker task force working on Standard P1453 is set to adopt the IEC standard as its own.

1.6.18. Standards Related to Distributed Generation

The new IEEE Standard P1547 provides guidelines for interconnecting distributed generation with the power system. This standard has been on a fast track and should be balloted shortly. It specifies requirements for the full range of interconnection issues and references appropriate standards for individual power quality issues where appropriate.

Many researches are going on to develop the standards for different problems occur in power quality research. They are following :

(*i*) **Harmonics and other low-frequency disturbances** : This area focuses on limits and methods of measurement for harmonics and inter-harmonics.

(*ii*) **Voltage fluctuations (flicker) and other low-frequency disturbances :** This area focuses on setting limits for voltage fluctuations caused by customer equipment and methods of appropriate measurement.

(*iii*) **Low-frequency immunity tests :** This area focuses on testing procedures that evaluate equipment immunity from power quality variations.

(*iv*) **Electromagnetic interference related to the network frequency :** This area is focusing the full range of power quality phenomena on the network and the interaction issues associated with end users of electricity.

(*v*) **Power quality measurement methods :** This area is currently focusing on IEC 61000-4-30, an overall guide defining the requirements for power quality monitoring equipment.

It's difficult to keep up with the wide variety of power quality standards and guides under development. The best way to stay on top of things is to participate in the process. As power quality professionals, we have the opportunity to increase our participation in these working groups and better coordinate the efforts of the IEC and IEEE.

Table 4 : Brief Description about Various Power Quality Standards

Standards	*Short Description of Standards*
IEEE 644	Standard Procedure for Measurement of Power Frequency Electric and Magnetic Fields from A.C. Power Lines
IEEE C63.12	Recommended Practice for Electromagnetic Compatibility Limits
IEEE 518	Guide for the Installation of Electrical Equipment to Minimize Electrical Noise Inputs to Controllers from External Sources
IEEE 519	Recommended Practices and Requirements for Harmonic Control in Electrical Power Systems
IEEE 1100	Recommended Practice for Powering and Grounding Sensitive Electronic Equipment
IEEE 1159	Recommended Practice for Monitoring Electric Power Quality
IEEE 141	Recommended Practice for Electric Power Distribution for Industrial Plants
IEEE 142	Recommended Practice for Grounding of Industrial and Commercial Power Systems
IEEE 241	Recommended Practice for Electric Power Systems in Commercial Buildings
IEEE 602	Recommended Practice for Electric Systems in Health Care Facilities
IEEE 902	Guide for Maintenance, Operation and Safety of Industrial and Commercial Power Systems
IEEE C57.110	Recommended Practice for Establishing Transformer Capability when Supplying Non-sinusoidal Load

IEEE P1433	Power Quality Definitions
IEEE P1453	Voltage Flicker
IEEE P1564	Voltage Sag Indices
IEC/TR3 61000-2-1	Electromagnetic Compatibility — Environment
IEC/TR3 61000-3-6	Electromagnetic Compatibility — Limits
IEC 61000-4-7	Electromagnetic Compatibility — Testing and Measurement Techniques — General Guides on Harmonics and Inter-harmonics Measurements and Instrumentation
IEC 61642	Industrial A.C. Networks Affected by Harmonics — Application of Filters and Shunt Capacitors
IEC SC77A	Low Frequency EMC Phenomena
IEC TC77/WG1	Terminology
IEC SC77A/WG1	Harmonics and Other Low Frequency Disturbances
IEC SC77A/WG6	Low Frequency Immunity Tests
IEC SC77A/WG2	Voltage Fluctuations and Other Low Frequency Disturbances
IEC SC77A/WG8	Electromagnetic Interference Related to the Network Frequency
IEC SC77A/WG9	Power Quality Measurement Methods

1.6.19. Standards Related to Custom Power

IEEE Standard P1409 is currently developing an application guide for custom power technologies to provide enhanced power quality on the distribution system. This is an important area for many utilities that may want to offer enhanced power quality services.

1.7. POWER QUALITY MONITORING

If an installation or network is thought to be suffering from problems related to Power Quality, suitable measurements should to be taken to confirm the initial diagnosis. These measurements will also help quantify the extent of the problem(s) and provide assistance in determining the most suitable solutions. Finally, follow up measurements after installation will confirm the effectiveness of the remedial measures taken.

1.7.1. Type of Installation

Monitoring equipment for Power Quality may be suitable for either temporary or permanent installation on a supply network. Permanent installation is most likely to be used by Utilities for routine monitoring of parts of their networks to ensure that regulatory limits are being complied with and to monitor general trends in respect of power quality issues. Consumers with sensitive loads may also install permanent monitoring devices in order to monitor Power Quality and provide supporting evidence in the event of a claim for compensation being made against the supplier if loss occurs due to a power quality problem whose source is in the Utility network.

The performance of any devices installed to improve Power Quality can also be monitored. Such devices may have a data link to a DCS or data logger in order to provide historical data recording and data processing/presentation facilities. They are quite small and are fitted in a suitable cubicle forming part of a switchboard line-up. The data link may be hard-wired, use a modem connection to a telephone line, or in the case of a utility with many geographically-dispersed substations, radio links for data transmission may be used. Internal data storage will be provided to ensure effective use of the data link. The units may be self- or auxiliary supply powered, and in the case of important Utility substations may have battery-backed supplies to ensure capture of voltage interruptions. Time synchronization may be required to ensure accurate identification of events.

For investigation of particular problems, a portable instrument is more suitable. The same range of Power Quality measurement capabilities is provided as for permanent instrumentation. The instrument may have built-in analysis/data storage capabilities, but external storage in the form of floppy discs or a data link to a laptop or desktop PC is commonplace. Analysis/report writing software running on a PC is often available, which may be more comprehensive than that provide in the instrument itself.

1.7.2. Connection to the Supply

Connection to the supply being monitored may present problems. For LV supplies, the voltage inputs are usually taken directly to the instrument in single-phase or three phase form as required. Monitoring of currents may be through a current shunt or suitable CT, depending on circuit rating. At higher voltages, VT's and CT are already fitted for instrumentation/protection purposes are used. In general, the conventional electromagnetic voltage or current transformer is suitable for use without special considerations being required, but capacitor voltage transformers often have a low-pass filter on the output that has the potential to seriously affect readings of harmonics and transient phenomena. In such cases, the input to the monitoring device must be taken prior to filtering, or the filter characteristics must be determined and the measured signals processed to take account of the filtering prior to analysis being undertaken. In addition, the CVT itself may have a non-linear transfer function with respect to frequency, though the variety of types of CVT and difficulties of testing make confirmation of this point virtually impossible at present.

Where harmonics or high-frequency phenomena are being measured, suitable connecting leads between the transducers and the measuring instrument are required to avoid signal distortion. This is especially important if long cable runs are used; this may be the case if the measuring instruments are centralized but measurements are being made at a number of switchboards.

1.7.3. Types of Power Quality Measurements

Instruments for power quality monitoring may not offer the full range of measurements for all Power Quality issues. Care is therefore required that the instrument chosen is suited for the purpose. Most instruments will provide measurements of current and voltage harmonics, and capture of voltage dips and frequency excursions.

Measurements to the commonly encountered standards may be built-in. For capture of surges, spikes and interruptions, more specialized instrumentation may be required as transient high-speed waveform capture is required. This requires a high sampling rate and large memory storage.

Most instruments designed for Power Quality use A/D conversion of the input waveforms. The raw waveform is stored and either transferred to a computer for analysis, or the instrument contains built-in software to carry out analysis of power quality in line with accepted standards.

Often the software will have a choice of standards for user selection. More detailed analysis using the same instrument can show directly how the results compare with this standard.

To facilitate the interchange of data between locations and/or users, the public-domain PQDIF data interchange format for Power Quality may be used and facilities provided for in the software.

1.7.4. Instrument Location

The location of the measuring instrument also requires consideration. By careful placement and observing the relative polarities, it is possible to deduce if the source of the disturbance is on the source or load side of the monitoring device.

SUMMARY

- *Harmonic distortion.* Continuous or sporadic distortions of the 60-hertz (Hz) voltage sine waveform, usually caused by microprocessor based loads in the building such as computer power supplies, lighting ballasts, and electronic adjustable speed drives. Harmonics can also be transmitted from an energy user down the block. These can cause telecommunications or computer interference; overheating in motors, transformers, or neutral conductors; decreased motor performance; deterioration of power-factor correction capacitors; or erratic operation of breakers, fuses and relays.
- *Interruption, momentary.* A very short loss of utility power that lasts upto 2 seconds, usually caused by the utility switching operations to isolate a nearby electrical problem.
- *Interruption, temporary.* A loss of utility power lasting from 2 seconds to 2 minutes, caused by a nearby short circuit due to something like animals, wet insulators, or accidents. Corrected by automated utility switching.
- *Long-term outage.* A loss of utility power lasting more than 2 minutes due to major local, area, or regional electrical events.
- *Noise.* Sporadic voltage changes consisting of frequencies higher than the normal 60-Hz power frequency due to any number of causes including arc welders, loose wiring, and nearby radio and TV transmitters.
- *Sag.* A short-term decrease in voltage lasting anywhere from milliseconds upto a few seconds. Sags starve a machine of the electricity it needs to function, causing computer crashes or equipment lock-ups. Usually caused by equipment start-up—such as elevators, heating and air-conditioning equipment, compressors, and copy machines—or nearby short circuits on the utility system.

- *Spike.* A very brief (nanoseconds to milliseconds) change in voltage ranging from tens to thousands of volts. Can be produced by utility and customer equipment operations, nearby lightning strikes, falling tree limbs on power lines, and even static discharges
- *Surge.* A short-term increase in voltage, lasting upto a few seconds. They are due either to customer equipment operation, such as air conditioners or motors switching on and off, or to utility activities, such as capacitor switching.
- *Transient.* A sudden momentary change in voltage. Also called a spike.
- *Undervoltage.* A decrease in voltage lasting longer than a few seconds. Usually due to undersized wiring at the facility but can also be caused by overloaded utility circuits and result in brownouts.

Review Questions and Exercises

1. Define Power Quality and discuss various power quality issues.
2. Write a short note on power quality Vs equipment immunity.
3. What are the standards given related to power quality ? Describe them.
4. Describe the various stages in power quality monitoring.
5. (*a*) Discuss various issues of power quality. *(MDU, Rohtak, May 2008)*

   (*b*) Describe electric power quality standards.
6. (*a*) Define power quality. Explain its various issues. *(MDU, Rohtak, Dec. 2008)*

   (*b*) Discuss power quality v/s equipment immunity.
7. (*a*) Define power quality. Explain various standards of electrical power quality. *(MDU, Rohtak, May 2009)*

   (*b*) Discuss power quality v/s equipment immunity.
8. Explain the operation of dynamic voltage restorer for correction of voltage sag/swell. *(MDU, Rohtak, May 2009)*
9. Write a short note on power quality monitoring. *(MDU, Rohtak, May 2009)*

❑❑❑

Chapter **2** # POWER FREQUENCY DISTURBANCES

2.1. INTRODUCTION

The major disturbances that are introduced into the power distribution systems either from the load centre or the utility including environmental causes, those can be classified into : steady state and transients. For steady state, the main disturbance is harmonics. For transients, the main disturbances are : sags, dips, momentary blackouts, surges, spikes, over/under voltages, frequency variations, power lines noise (common and normal mode interference), conducted radio frequency and electromagnetic interference.

2.2. COMMON POWER FREQUENCY DISTURBANCES

Common power quality disturbances include transients, voltage swell and sags in power source voltage and harmonics (or "noise") on the power line. Each of these occurrences is discussed briefly below.

2.2.1. Voltage Swell

It is a rapid short-term increase in voltage. Swells often are caused when high power demand devices such as air conditioners turn off and the extra voltage is dissipated through the power

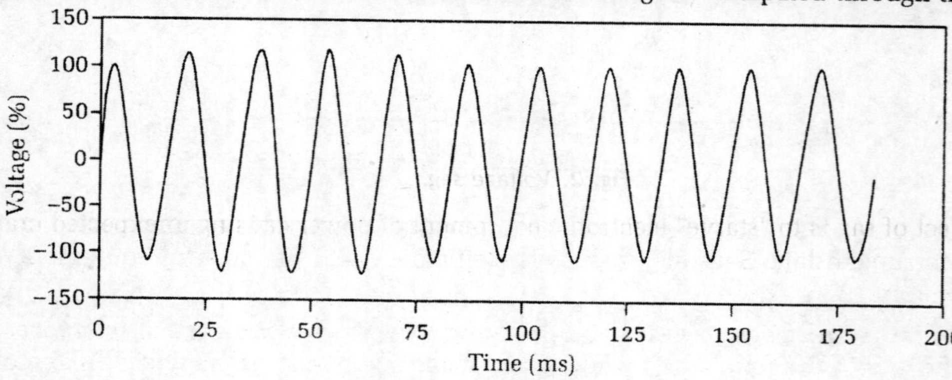

Fig. 1. *Voltage swell.*

line. Since sensitive electronic devices require a constant voltage, swells stress delicate components and cause premature failure.

2.2.1.1. Swell Causes

As discussed previously, swells are less common than voltage sags, but also usually associated with system fault conditions. A swell can occur due to a single line-to ground fault on the system, which can also result in a temporary voltage rise on the unfaulted phases. This is especially true in ungrounded or floating ground delta systems, where the sudden change in ground reference result in a voltage rise on the ungrounded phases. On an ungrounded system, the line-to ground voltages on the ungrounded phases will be 1.73 p.u during a fault condition. Close to the substation on a grounded system, there will be no voltage rise on unfaulted phases because the substation transformer is usually connected delta-wye, providing a low impedance path for the fault current.

Swells can also be generated by sudden load decreases. The abrupt interruption of current can generate a large voltage, per the formula : $v = L\, di/dt$, where L is the inductance of the line, and di/dt is the change in current flow. Switching on a large capacitor bank can also cause a swell, though it more often causes an oscillatory transient.

2.2.2. Voltage Sag

It is a rapid short-term decrease in voltage. Sag typically is caused by simultaneous high power demand of many electrical devices such as motors, compressors and so on.

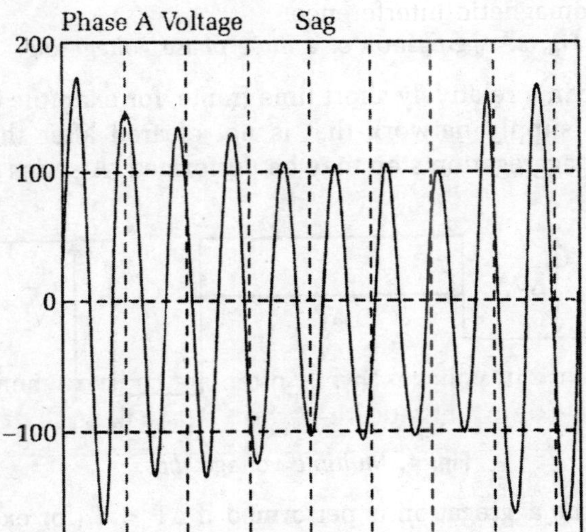

Fig. 2. *Voltage sag.*

The effect of sag is to "starve" electronic equipment of power causing unexpected crashes and ost or corrupted data. Sags also reduce the efficiency and life span of equipment such as electric motors.

Sags are drops in rms voltage for less than few seconds, beyond normal tolerance. The causes may be any of the following : when heavy loads are started or energized (like motors, air conditioners, large inductive loads) or when flashover of lines occurs due to lightning, when power system faults occur.

2.2.2.1. *Aggregation of Voltage Dips*

A voltage dip may refer to one phase, two or three phases. A three phase voltage dip can be symmetrical or asymmetrical. Aggregation of voltage dips is sometimes performed in order to represent it according to two parameters : duration and depth.

A method of three phase voltage dip aggregation is presented in Fig. 3. The dip starts when the voltage of the first phase drops bellow a certain point and ends when the last phase recovers (in Fig. 3, $\Delta T = t_6 - t_1$). The dip depth is defined as the maximum of the three values of the different phases (in Fig. 3, $\Delta V = \max[\Delta V_1, \Delta V_2, \Delta V_3]$)

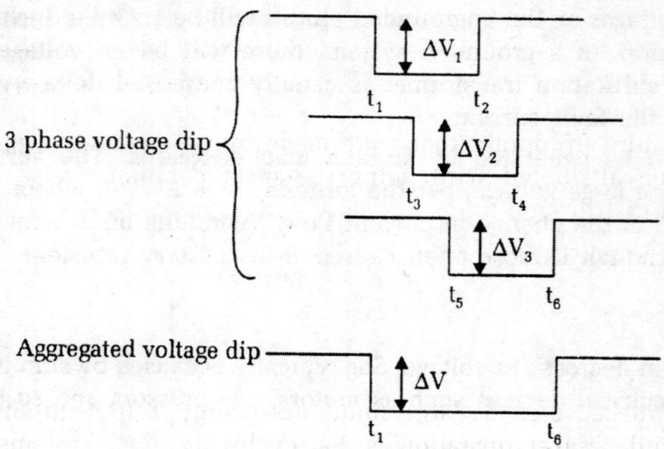

Fig. 3. *Aggregation of a three phase voltage dip.*

If several dips occur in a relatively short time frame, for example two dips that are caused by a short-circuit in the supply network that is not cleared after the first re-closing of the current breaker, a second aggregation step may be performed. A series of two dips is presented in Fig. 4.

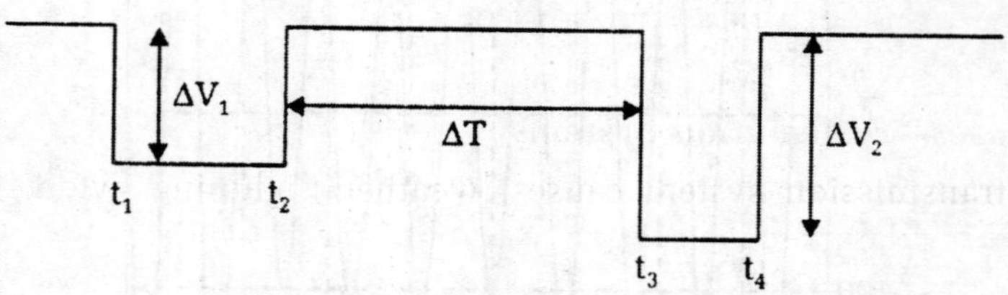

Fig. 4. *Multiple voltage dips.*

The second step of the aggregation is performed if $\Delta T < T$ (for example, $T = 1$ sec.). The parameters that describe the aggregated event are : depth $\Delta V = \max[\Delta V_1, \Delta V_2]$ and duration :

(1) $\Delta t = t_4 - t_1$,

(2) $\Delta t = \max\,[(t_2 - t_1)\ (t_4 - t_3)]$,

(3) $\Delta t = (t_2 - t_1) + (t_4 - t_3)$.

2.2.2.2. *Causes*

The causes of sags have been broken into three areas of occurrence : the transmission system (typically above 65 kV); distribution systems (65 kV to 12 kV); and, point-of utilization (120–480 V). Swells are treated under a single category. A common, underlying cause of sags and swells in all three areas is a sudden change of current flow through the source impedance.

An understanding of Ohm's Law and Kirchhoff's Equations, as they relate to real-life, non-ideal sources, is necessary to understanding the effects of such. In the case of a sag, the sudden, large increase in the current required from a source will cause a larger voltage to be developed across the source impedance. This will result in a reduction in the voltage, as seen by the load. Likewise with a surge, a sudden reduction in the current flow will cause an increase in voltage in inductive/capacitive impedances, which the load may experience.

2.2.2.3. Sag Causes—Transmission Systems

The causes of voltage sags on a transmission level system are similar to those on a distribution system. They include the weather (especially lightning), construction accidents, transportation accidents (helicopter or light planes are common culprits), animals or a fault on another part of the system causing "sympathetic" sags. There have been recorded instances of the nesting habits of large birds in the towers resulting in phase-to-ground faults when the insulators were "shorted out" by bird droppings that were made into a conductive path during rain storms. Lightning is often attributed with being the most common cause of faults on overhead transmission and distribution lines. The fault can occur by lightning directly striking a phase conductor, or by striking a grounded object, such as shield wire or tower, which is called a backflash. A flashover develops from the voltage path across the phase conductors to ground or to another other phase, resulting in flow of fault current.

Transmission-related voltage sags are normally shorter in duration than distribution voltage sags. This is attributed to the fact that the fault clearing mechanisms (the relay/breaker schemes) must react faster, because of the large amount of energy in transmission faults. Total time for fault detection and breaker operation is 3–6 cycles on older systems, with newer breakers having fault clearing times within a cycle.

Another reason for the shorter duration is that transmission systems are looped or networked, versus radial for distribution systems. This means that when a single line trips, the remaining system can still handle the load, including the fault current. However, the larger currents involved can have further reaching effects. The effects of sag to 90% have been found to be experienced up to 1000 kilo meters away from the fault, while a sag to 75% have effect up to 500 kilometers away.

2.2.2.4. Sag Causes—Distribution Systems

Similar to the transmission system causes, weather (lightning, wind, ice), animal contact, contamination of insulators, construction accidents, motor vehicle accidents, falling or contact with tree limbs can result in voltage sags. Such faults may be 3-phase, line-to-line, or single line-to-ground. The 3-phase faults are the most severe, but are relatively unusual. "Single line-to-ground faults on the utility system are the most common cause of voltage sags in an industrial plant". Preliminary results from the many studies indicate that most important cause of momentary voltage sags is lightning strikes. In the majority of sags, the voltage drops to about 80% of nominal value on the parallel feeders, while the faulted feeder may have a lower sag value, or may result in an outage if the fault cannot be cleared.

Distribution system sags tend to cluster around several duration ranges based on the fault protection schemes : 6–20 cycles (typical distribution fault clearing times, 30–60 cycles (the instantaneous reclosing time for breakers) or 120–600 cycles (the delayed reclosing time for breakers).

2.2.2.5. Sag Causes—Point of Utilization (Loads)

In a study, 50% or more of the recorded low/high RMS events were caused by load equipment in the same building. Sudden increases in the current requirement can have the same effect within a facility's wiring as on a utility distribution system. Voltage sags can be caused by fault conditions within the building, or the start up of large inductive loads, such as motors, that create a temporary in rush current condition. The starting of large horsepower motors that would draw adequate current are typically longer in duration than 30 cycles, and the associated voltage magnitudes are not as low as with a utility fault. The voltage sag condition lasts until the large current demand decreases, or the fault is cleared by a protective device. In the plant, this will typically be a fuse or a plant feeder breaker.

2.2.3. Transient Disturbances

Transient power problems originate outside buildings from ground faults, lightning, public utility switching. They also originate inside buildings from inductive loads and switching. Photocopiers, air conditioners, compressors, motors, light switches and fluorescent lighting circuits are all common sources of power line transient disturbances. Any collapsing magnetic field or any sparking in a switch will create transient waves which will appear at various locations with different modes and the consequences will vary.

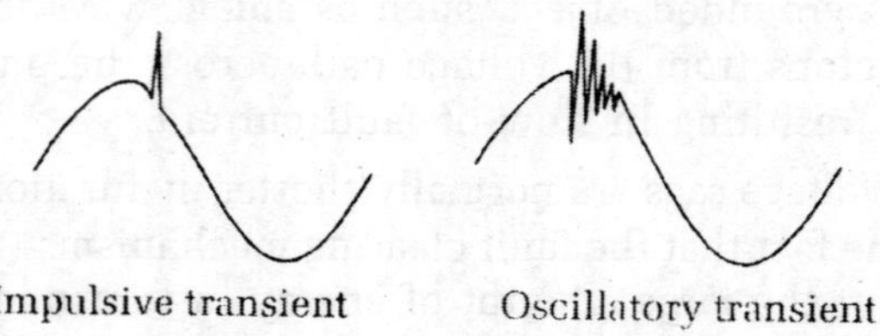

Fig. 5. *Transients in a signal.*

It is an extremely high and nearly instantaneous increase in voltage with a very short duration measured in microseconds. Transients are often caused by lightning or by events such as power coming back on after an outage. Transients can damage or destroy sensitive electronic equipment. Turn the equipment off during a power·outage. Wait a few minutes after power is restored before turning it on, and then turn on one device at a time.

2.2.3.1. The Effects of Transient Disturbances on Loads

As can be seen from the above, the disturbances that any load may see can vary from under-voltage to over-voltage, to surges, to noise and interference. Certain disturbances can cause overheating and if lasted can cause failure of equipment, example motors when operated with voltage below nominal. Other loads like heaters will overheat with over voltage and can burn itself out if the disturbance occurs frequently or last for long duration. With spikes electronic loads can be destroyed and motor or transformer insulation may breakdown. Noise disturbs microprocessor based equipment (*e.g.*, programmable logic controllers and microcomputers) and it may blow power supplies and boards, burn chips, cause memory loss, premature component failure, data error and unexplained microprocessor malfunction. Electromagnetic interference can lead to malfunction of supervisory control equipment, inverters, digital metering equipment and industrial controllers.

2.2.4. Noise

It can be described as a disturbance in the smooth flow of electricity. It is often technically referred to as electro-magnetic interference (EMI) or radio frequency interference (RFI). "Harmonics" are a special category of power line noise that causes distortions in electrical voltage. Noise can be caused by motors and electronic devices in the immediate vicinity or far away. Noise can affect performance of some equipment and introduce glitches and errors into software programs and data files.

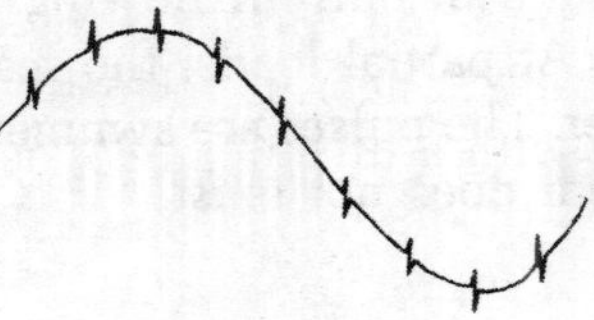

Fig. 6. *Noise in a signal.*

It can be classified into normal (differential) and common mode noise. The normal mode is noise that appears between the hot and the neutral wires, common mode is between the hot or neutral and true ground. Microprocessor and digital logic control systems utilize the voltage between neutral and ground (theoretical zero) as a zero voltage reference. These distortions of the normal sine wave can be caused by transmitters, fluorescent lights, and electronic control circuits.

2.2.5. Interruptions

It is the total loss of power for some period of time. Interruptions are caused by excessive demands on the power system, lightning strikes and accidental damage to power lines. In addition to shutting down all types of electrical equipment, Interruptions cause unexpected data loss. **(See Chapter-1)**

> ***Note :*** *For other common power frequency disturbances see Chapter-1 under heading-Power Quality Issues.*

2.3. SOURCES OF STEADY STATE DISTURBANCES

The sources of steady state disturbances like harmonics are arc furnaces, welders, saturable reactors, rectifiers, thyristor controlled loads, variable voltage/frequency drives, switching mode power supplies, UPS, magnetizing currents. Voltage notching is also produced by certain devices aforementioned like adjustable frequency drives.

2.3.1. Arc Furnaces and Welders

When the arc is stable the non-linear resistance characteristics of the arc tends to cause a peaked current wave and a flattop voltage. The combination of the non linear arc resistance and large linear external resistance results in more apparent distortion to the supply voltage than to furnace current. The peaked current wave will have harmonics comparable to square waves but with different phase relations. A peaked or triangular wave consists of odd order harmonics. The 3rd, 7th, 11th etc., harmonics cross the zero reference line 180° out of phase: 5th, 9th, 13th etc., cross the reference in phase with the fundamental. In the square wave, all the harmonics are odd order and cross the reference line in phase with the fundamental. When the arc within the arc furnace is unstable during the meltdown period, sudden changes in arc length would cause sudden changes of current magnitude, which in turn would cause voltage flicker.

2.3.2. Saturable Reactors

These devices control the magnitude of current flowing to a load by means of varying the reactance of the saturable reactor. At partial loads, the resultant current wave is very similar to that of a phase controlled rectifier. The pulses are symmetric and will contain odd harmonics. At full load the harmonics problem does not exist.

2.3.3. Rectifiers

A broad classification of rectifiers is whether it is half wave or full wave, a further classification is the number of pulses whether 3, 6, 12, or 24. The harmonics available on a line feeding rectifiers are obtained from *pn* +/1; where *p* is number of pulses and *n* is a series of numbers : 1, 2, 3, 4, 5, 6... A full wave rectifier with a smoothing choke in series with the load will produce on the a.c. side a square wave of equal positive and negative values. This square wave is symmetric and will contain the fundamental with all the odd harmonics in phase with the fundamental with magnitudes, approximately, inversely proportional to their order. Other common configuration using three phase input with a Wye-Delta rectifier transformer will have no third harmonics or its multiples, it is known as 6 pulse rectifier. Some rectifier circuits are provided with smoothing capacitors (in parallel with load) in place of a choke. In this case, the capacitor will supply energy to hold up the voltage across the load. The current will flow only during the period of time that the supply voltage is greater then the voltage across the capacitor. Thus pauses will exist between the positive and negative current pulses. The harmonics generated in this case will be of the odd order.

2.3.4. Thyristor Controlled Loads

Phase controlled devices will vary the relative magnitudes of the harmonic components as they are adjusted to control the supply to the load. A phase controlled device with 90°, delay may generate a wave similar to the saturable reactor at partial load. With equal phase control in both the positive and negative half waves, the waves are symmetrical and will, only, contain odd harmonics. If the control of each wave is independent (full wave rectifier with a phase controlled device in only one leg and a diode in the other), the presence of even harmonics become a possibility. The phase relationship between the fundamental and the harmonics vary with the phase control setting of the thyristor.

2.3.5. Variable Voltage/Frequency Drives

In general each drive by itself does not present a problem, but with the application of many drives, the problem of harmonics and voltage notching may materialize. Line voltage notching and line harmonics currents result into line voltage distortion. With variable frequency drives the static power converter which is the front end of the adjustable frequency is the source of line disturbance. This converter may be a three phase full wave bridge rectifier using either diodes or SCRs. The shape of the distorted line current wave from the adjustable frequency drive will be determined by the type of converter, load and the impedance of the power source. For an inductive load, the current wave will be close to a square wave, it is typical for SCR converters, six pulse and current source drives without D.C. chopper. For a capacitive load, the current wave will be non-continuous, symmetric, twin half sine waves, typical for diode converters, pulse width modulation drives and six pulse drives with D.C. chopper. The final magnitude of the various line harmonic currents will be determined by the inductance in the power system commutating reactance, the load on the drive and if SCR are used, on the firing angle (delay angle) of the thyristor.

The total harmonic current distortion that a drive produces on the A.C. line is given by :

$$= (\text{sum of square of individual harmonic currents/ fundamental current})^{0.5}$$

The above expression can be applied to any harmonics generating device. Voltage notching only occurs when drives produce continuous square current waves but not with non-continuous current. The notches occur when continuous line current transfers (commutates) from one phase to another in a static power converter. During commutation, two phases are connected for a very short duration through the A.C. source impedance (very low) by the converter; this will cause the voltage to drop near zero.

The voltage notching is expressed by the following :

$$= (\text{sum of square of all harmonics voltages/fundamental voltage})^{0.5}$$

This is equal to the voltage distortion factor. This factor is related to the area and depth of the voltage notch and the reactance of the circuit. In some cases, when SCRs are used in the converter, voltage spikes of very short duration can be generated immediately following the notch. It is a phenomenon associated with solid state devices.

2.3.6. Switching Mode Power Supplies

D.C. power supplies, +/5V or +/12V for computers and other microprocessor based electronic equipment is a major source of harmonic distortion and noise.

The interface between the mains network and any electronic equipment is a switch mode power supply (SMPS). Every electrical device including electronic control, require a power supply for the polarization of the electronic devices. Even in power electronics systems, the control circuitry is fed by an auxiliary SMPS.Consequently, the failure analysis of electronic circuits, due to power quality problems, implies the knowledge of the SMPS behaviour, under abnormal situations of the AC input. The Fig. 7 shows the basic building blocks of a SMPS. The input filter is intended to suppress radio frequency noise that results from transistor switching transitions in the D.C.-A.C. square wave converter.

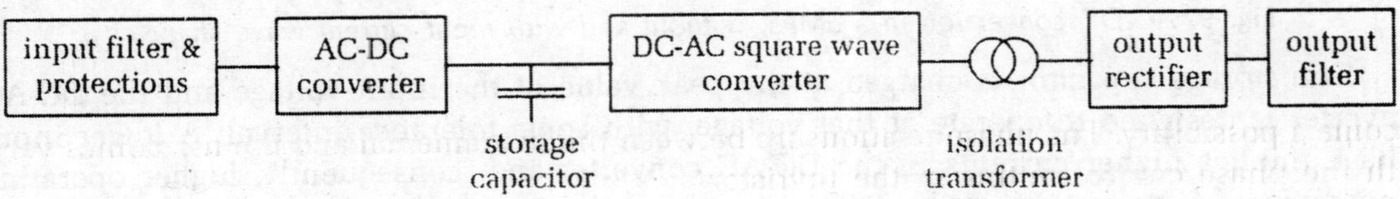

Fig. 7. *Building blocks of a switch mode power supply.*

Typically, it presents an attenuation of 30 to 50 dB, at frequencies above 1 to 5 MHz and is designed to operate with noise amplitudes up to 1 to 5mV. This filter is not efficient in the limitation of high amplitude, or high energy transients. The capacitances C_X and C_Y are very small (<100pF) and are designed to present a low impedance at frequencies higher than 1MHz. The input protection of SMPS includes usually an over-voltage protection, as shown in Fig. 8.

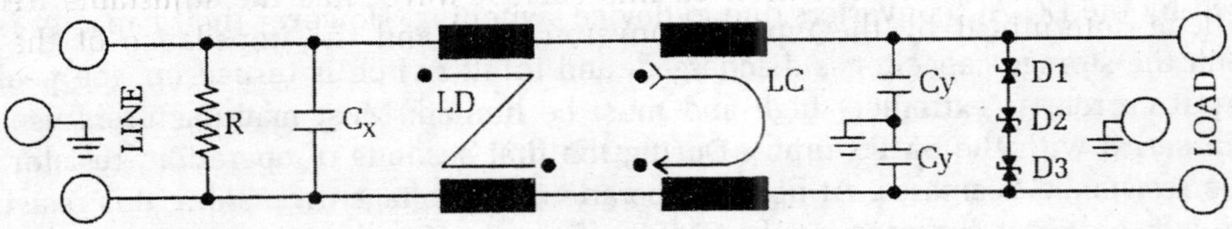

Fig. 8. *Radio frequency input filter of switch mode power supplies.*

In this case, a group of 3 transient suppressors in series is mounted at the output of the filters. Those suppressors are in parallel with the line input and can afford an energy dissipation of less than 10J. An high energy transient blows the suppressors and, afterwards, damage the converters. The time response of those devices is quite low (<10ns), but it is not sufficiently low to provide an accurate protection of sensitive devices, like power MOSFETS. In conclusion, the input filter of a SMPS is designed with a single objective-compliance with electromagnetic compatibility standards-and does not provide any protection of the power supply against the line input disturbances.

Some SMPS manufacturers introduce, at the system input, an overvoltage protection that blows the input fuse, when the input voltage remain higher than 110% up to 130% of the nominal value. Those protections include a voltage detection circuit and a thyristor, and are very slow (10–100 μs).

The AC/DC conversion is usually obtained with a diode bridge and capacitive filtering. The output voltage is approximately equal to the peak voltage of the mains input. The capacitor is a bulky storage device, due to two different design objectives :

- reduction of the residual AC component of the voltage (few volts peak to peak);
- energy storage that permits the operation of the power supply, during a short interruption of the mains supply, even when it works at full power; the allowed interruption time is typically 30ms up to 60ms (standon time), without any variation of the output voltage.

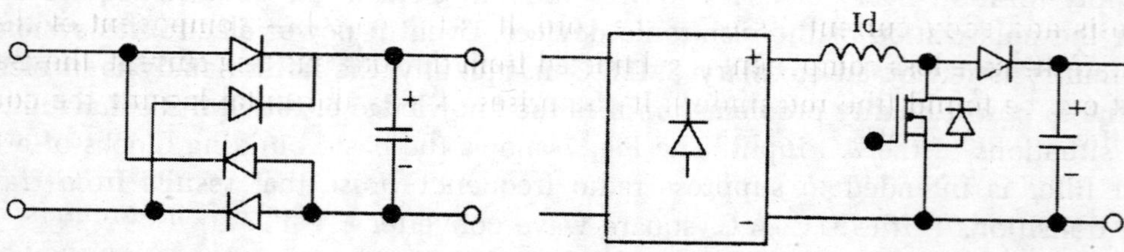

Fig. 9. *AC-DC conversion in a SMPS, without and with input current wave shape.*

The storage capacitor is charged at the peak value of the input voltage and the DC-AC converter is designed to operate at that voltage, with some tolerance interval. A lower input voltage implies higher currents in the DC-AC converter and, consequently, higher operating temperatures. A correct rms value of the input voltage is not sufficient to avoid the SMPS overheating. Very often, the rms value is acceptable and the peak value of the voltage is much lower than V_{rms}, due to the effect of voltage distortion (3rd, 5th and 7th harmonics).

The storage capacitors are large electrolytic capacitors, performing a good filtering effect at 100Hz. Their equivalent series resistance (ESR) and equivalent series inductance (ESL) are very large. Consequently their impedance rises for frequencies above some few kHz (1–10 kHz), resulting in a very poor filtering effect at higher frequencies. Additional capacitors, with a better frequency response, are usually connected in parallel, to avoid an excessive noise introduced by the DC-AC converters due to device switching. However their capacity is small.

When the storage capacitor is discharged, and input switch is turned on, the peak value of the input current is extremely high and must be limited. Most manufacturers use a NTC resistor in series with the SMPS input. During the first seconds of operation, resistor warms up and its resistance decreases. At normal operation, the voltage drop along this resistance is small enough to do not compromise the SMPS efficiency. The thermal time constant of those resistances is usually 5–10s. Consequently, the occurrence of a power failure with a restart in less than 10s is dangerous to the integrity of the input rectifier, because the NTC resistor is hot and does not limit the inrush current and input fuse may be not sufficiently fast to avoid the effects of the over-current.

2.3.7. Uninterruptible Power Supply (UPS)

The output distortion of a UPS depends upon its design and the output impedance. Usually the distortion caused by the UPS is specified by the manufacturer. It is given as a percentage harmonic distortion and is limited to 5% total harmonic distortion. This number is based on linear loads or reactive inductive loads. In cases where the UPS supplies nonlinear loads, the harmonic distortion can be higher than with linear loads. (Details are later in this chapter)

2.3.8. Magnetizing Currents

When transformers are energized, inrush currents that are rich in second harmonics will be seen by the primary of the transformer. The duration of such phenomenon can vary from 10 cycles to 3600 cycles (1 minute) function of the size of the transformer bank. After the elapse of the inrush period the exciting current which is typically 5% of the full load current of the transformer will continue to appear in the primary of the transformer and is rich in the third and fifth harmonics. If the exciting current is analyzed by Fourier series method, it will be found to comprise a fundamental and a family of odd harmonics. The fundamental can further be resolved into two components, one in phase with the counter EMF and the other lagging by 90 degree The fundamental in phase component accounts for the power absorbed by the hysteresis and eddy current losses in the core. It is the core loss component of the exciting current. If the core loss component is subtracted from the total exciting current, the magnetizing current can be found (the remainder). It comprises of a fundamental lagging the counter emf by 90 degree and all the harmonics. For a typical power transformer, the third harmonics is 40% of the exciting current. Briefly, arcing faults and their effect on power quality will be discussed. In an arcing fault (to ground) on low voltage systems discontinuous current is produced. The reason for the discontinuity is the period of time before which the voltage arcs over from the line to ground. It will extinguish at the first current zero and then reignite on the succeeding wave of opposite polarity. The wave is symmetric and will have considerable odd harmonics content.

2.4. DISTURBANCE RECOGNITION SYSTEM

In order to get rid of visual inspection of the waveforms, an automated method is required. This will involve the integration of signal processing and artificial intelligence techniques like neural networks, fuzzy logic. A new signal processing tool *i.e.*, Wavelet transforms is also added to the arsenal of automatic power quality disturbance recognition. The artificial intelligence techniques are used for decision making. This integration gives rise to an automatic disturbance recognition system (ADRS).

The ADRS will receive data online, process them and will be able to determine whether there is a PQ disturbance in the particular group of data. In the event that there is a disturbance, the system will also be able to classify the type of disturbance. The components of the ADRS are shown in Fig. 10.

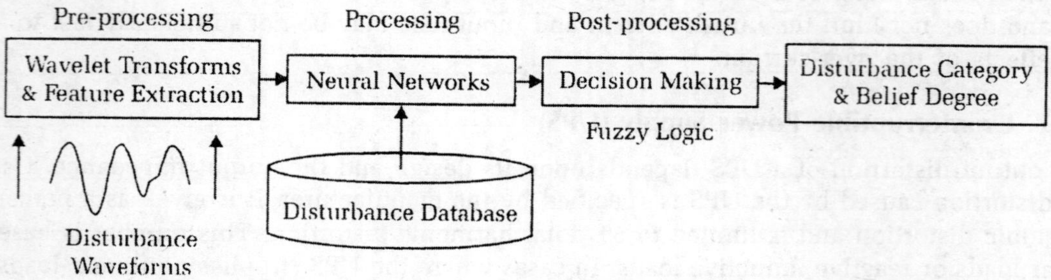

Fig. 10. *Block diagram of the automatic disturbance recognition system.*

2.4.1. Modelling for the ADRS

In order to train the Automatic Disturbance Recognition System, a large database of disturbance waveforms is required. There are two possible sources for these waveforms. The preferable source of disturbance data is from recording actual disturbances in the power system. However, it is clear that only a small percentage of possible disturbances will occur in the power system over any given period of time. This means that the source of the majority of disturbance data is from simulation.

The source of a recorded disturbance also has to be identified so that the recognition system can be trained to identify similar disturbances as coming from that source. If the source of the disturbance is not obvious, then it can be compared to similar simulated disturbances to locate the source.

Suitable models must be constructed for each category of disturbance the ADRS is to recognize. These include load flow and fault analysis for long duration disturbances, harmonic models and higher frequency models for transient disturbances. In the majority of cases, only the local area around the disturbance source need be included into the model in detail.

Relatively detailed load models are a requirement for good power quality simulation. The possible disturbances that a load can create must be understood as well as a load's response to external disturbances. To date, a comprehensive 3-phase harmonic model of the Tasmanian high voltage transmission network has been constructed. The results produced by this model have been compared to a series of measurements that were taken and analyzed to determine the relative contribution of harmonics at different locations, from different sources. The system was tested with data containing known disturbances. The percentages of successful recognition of the various types of disturbances are listed in Table 1.

Table 1 : Success Rate of Disturbance Recognition

Disturbance Category	*Correct Recognition (%)*
Voltage Sag	99
Voltage Swell	95
Momentary Interruption	100
Voltage Fluctuation	93
Cap. Switch. Transients	93

Impulsive Transients	100
DC Offset	100
Frequency Deviation	95
Notch	95
Noise	59

It may be noticed in this table that for different types of disturbances, the success rate varies from 93 to 100 percent except for the noise. This may further be improved if the data is further processed. Work is presently going on towards this direction. The relatively poor success rate in the recognition of noise is due the low sampling frequency adopted in capturing the distorted waveforms. It is expected that with the increase of sampling rate, the success will also improve.

2.5. THE EFFECTS OF STEADY STATE DISTURBANCES ON LOADS

The degree to which harmonic currents and voltage notching (distortion) can be tolerated is function of the susceptibility of the type of load and other components of the power distribution system to such disturbances. The effects of steady state disturbances on different loads are as under :

2.5.1. Motors and Generators

The major effect of line disturbances on these machines is increased heating due to copper and iron losses at the harmonic frequencies. In some severe cases, pulsating or reduced torques may occur. The sum effect of the harmonics is a reduction in the efficiency and life of the machine. Generator regulating controls can also be affected by harmonic currents and voltage notching. It may result in hunting and instability and may make paralleling of generators very difficult.

2.5.2. Transformers

Voltage harmonics cause higher transformer voltage and insulation stress. Higher voltage will cause over excitation and consequently high harmonics contents in the line currents due to the higher excitation current. Transformer heating, increased copper and iron losses and reduced life are the results of high harmonic contents in the transformer phases (lines). If the system is 4wire, overheating of the neutral due to the flow of the zero sequence harmonic current is a possibility.

2.5.3. Power Cables

Medium voltage cables involved in system resonance (because of the inductance of lines or equipment and distributed or lumped capacitance) that is caused by harmonics current may be subjected to voltage stress and partial discharge that may lead to insulation failure.

The flow of harmonics currents in a conductor will cause additional heating due to skin and proximity effects both of which are function of frequency or order of harmonic, size and spacing of cable. In a four wire systems, additive harmonics in the neutral conductor can result in a neutral current exceeding the phase currents. This may lead to overheating of the neutral conductor that may affect its insulation and its surroundings. In 3wire systems, the third, ninth, etc., harmonics are not present.

2.5.4. Switchgear and Motor Control Centers

Excessive harmonics can increase heating and losses in switchgear/MCBs, reducing current carrying capability (continuous) of fuses and shortening the life of the insulating components.

2.5.5. Protective Relaying

For electromechanical relays additional torque components are produced and may alter the time delay characteristics of the relay. This may lead to incorrect tripping of the relay. For solid state, non RMS sensing, nuisance tripping or mal operation can result if the sensed current is rich in harmonics.

2.5.6. Metering and Instrumentation

These induction disk meters are affected by harmonic currents and voltages. The conditions can worsen if resonance results in high harmonic voltage. The induction disk devices see only fundamental current, harmonics can cause phase unbalances which can lead to mal operation. Positive or negative errors are possible depending on the harmonics involved and the design of the meter. Using the RMS sensing meters and relays will alleviate the aforementioned problems that are encountered with peak sensing and induction disk meters and relays.

2.5.7. Electronic Equipment

Power electronic equipment is susceptible to misoperation caused mostly by voltage notching. The voltage notch can distort the voltage wave form near zero crossing point, making it difficult to synchronize the firing pulses for phase controlled devices. Computers, electronic instruments and control devices may incur erratic operation and sometimes malfunction should the harmonic factor exceeds 5% and 3% for a single harmonic.

2.5.8. Circuit Breakers

The presence of harmonics, function of the type of the interrupting media and the arc quenching process, may cause the breaker to fail to interrupt the currents flowing through it. Under resonance conditions, the voltage appearing across the open circuit breaker may cause the failure of the breaker due to stressing the insulators and insulating medium.

2.5.9. Communications

When a long communication line (telephone line) is run in close proximity to a power line feeding solid state devices and nonlinear loads, harmonic interference can occur. The effect is interference with the transmission of information. In some cases, harmonics can emulate a signal and cause specious system responses.

2.5.10. Capacitors

Capacitors with the inductance of the system can cause resonance at a frequency other then the fundamental. This may impose a higher than rated voltage on the capacitor and may cause excessive currents to blow capacitor fuses. The life of the capacitor may be shorten because of the overheating due to current harmonics or the over stressing due to the voltage distortion. As a rule of thumb, the life of a capacitor will be halved for every 10°C temperature rise above normal or for 10% above rated voltage.

2.5.11. Other Devices

Ballasts for fluorescent or mercury lighting may have capacitors which together with the inductance of the ballast and the circuit have a resonant value. If this value corresponds to one of the generated harmonics, excessive heating and consequently failure of the fixture will occurs. Clock or paging systems can be adversely affected when one of the harmonic frequencies falls on or near the carrier frequency. Voltage notching can also cause misoperation of clock systems.

2.6. TECHNIQUES TO REDUCE DISTURBANCES

Before getting into the available options to minimize the adverse effects of harmonics, series and parallel resonance will be covered, briefly. Inductive reactance increases directly with frequency ($2\pi fL$) and capacitive reactance decreases directly with frequency ($1/2\pi fc$). At the resonant frequency, of any inductive/capacitive (*LC*) circuit, inductive reactance (X_L) equals capacitive reactance (X_C). For the series resonance circuit, the total impedance at the resonant frequency reduces to the resistance component only. High current magnitudes at the exciting frequency will flow. For the parallel resonance circuit, the total impedance is very high at the resonant frequency. High voltage magnitudes will appear across the load and high circulating current will flow in the capacitance-inductance loop, although the source current will be small in comparison.

2.6.1. The Available Techniques to Alleviate the Harmonics Problems

2.6.1.1. Shunt Filters

A number of shunt filters tuned to specific harmonic frequencies (5, 7, and 11) can effectively reduce the harmonics currents in a power system. Each filter consists of a series *LC* circuit tuned to resonate at a specific frequency and is located relatively near the harmonic source. It will provide a low impedance path and will shunt most of the harmonics current to ground thus preventing it from getting into the power system. In some cases these traps can overheat if they trap harmonics from other sources than those intended for.

2.6.1.2. Series Filters

These filters consist of a parallel *LC* circuit tuned to resonate at a specific frequency and are located at the equipment being affected.

2.6.1.3. Zero Sequence Filters

It is a specially designed zigzag reactor which provides a low impedance path for all zero sequence currents. This filter provides low impedance path for currents in phase and high impedance to currents not in phase (in case of 3 phase systems to currents, 120° apart).

2.6.1.4. Phase Multiplication

It involves phase shifting transformer to distribute power to harmonics generating devices, mainly rectifiers and converters. By properly shifting the phase relationship to various six pulse converter loads, the new effect in the power system is to create 12 or 24 pulse circuit. The result would be fewer harmonic generated and transmitted through the power system. As mentioned under rectifiers, above, the harmonics remaining in the power system will be np +/1.

2.6.1.5 Harmonic Injection

Harmonics currents can be eliminated by inducing harmonic fluxes in the core of a transformer with 180 degree phase shift from the harmonic fluxes induced by current flowing in the transformer secondary. These devices are designed to constantly monitor the load current, inject a current equal and opposite to the distorted component (thus cancelling it).

2.6.1.6. Power System Design

The effects of voltage distortion and harmonics currents can be minimized on other system loads by properly locating and isolating harmonics producing equipment as much as reasonable. The power wiring to a converter can be isolated from the control wiring or other load conductors to minimize the inductive and capacitive coupling between the different wiring systems. Voltage distortion can be minimized if the adjustable frequency drive or the harmonic producing devices load, as a percentage of total transformer load is kept to a minimum, also if the distribution transformer short circuit level to the actual load current ratio is increased. Major resonant conditions can be minimized by locating relatively smaller banks of power factor correction capacitors throughout a power system. These can be switched as needed versus using a single large bank of capacitors.

2.6.1.7. Reactors or Isolation Transformers

The effect of reactors or isolation transformers is to slightly change the shape of the current wave, thus lowering not eliminating the amplitude of the harmonics currents. The reactor is placed between the harmonics generating device and the point of common coupling. The distortion to the power system at the point of common coupling may be reduced.

2.6.1.8. Reducing System Impedance

As voltage drop in a power distribution system is function of each harmonic current through the impedance, voltage distortion can be reduced by either lowering the current or the impedance. Keeping the conductor lengths as short as practically possible, using low impendence transformers (this method has other advantages and disadvantages), using delta primary (this method has its disadvantages) winding in the distribution/rectifier transformer will eliminate the zero sequence currents path outside the delta, these are just a few ways to reduce the impedance in a power system.

2.6.2. Corrective Measures to Transient Line Disturbances

2.6.2.1. Built-in Filter

Most equipment has some power line filtration built into the power supply. These filters are primarily designed to stop disturbances entering from the equipment into the power line rather than the other way around.

2.6.2.2. Dedicated Lines/Isolated Grounds

These are separate lines running from a panel to the sensitive loads. The ground wire is insulated and runs directly back to the panel ground. Power disturbances should be considered as a likely problem even though a dedicated line has been installed.

2.6.2.3. *Regulation Transformers*

These are designed to maintain the rms voltage within certain levels. They will accept input voltages of plus or minus 15 percent from the rating and provide output voltages of 1 or 2% variance from the rating. They have no effect on transient disturbances or common mode noise.

2.6.2.4. *Isolation Transformers*

They incorporate shielding and grounding to eliminate common mode noise. They do no attenuate normal mode disturbances. If more than one load is connected to the output of the transformer ground loops will occur.

2.6.2.5. *Saturated Transformers*

These are designed to operate with the primary coil in full saturation. They eliminate normal mode disturbance but output a square wave. They can add further disturbances to the system like harmonics.

2.6.2.6. *Ferro-Resonant Transformers*

These are designed to operate in saturation with suitable capacitors on the output coils to create a resonant sine wave at 60 c/s.

2.6.2.7. *Ultra Isolation Ferro-Resonant Transformers*

They have the features of the isolation transformers and the Ferro-resonant transformer; probably they will give good protection against common and normal mode. They are load and frequency sensitive. They produce heat and audible noise.

2.6.2.8. *Line Conditioners*

They incorporate the features of regulation transformers and isolation transformers. Clamping devices are added to reduce spikes. They offer good regulation of rms voltage and good ground isolation from common mode noise.

2.6.2.9. *Clamping Devices*

These are also known as surge suppressers and sometimes surge arresters. These are non-linear voltage sensitive devices such as silicone carbide, gas discharge tubes and/or metal oxide varistors. These devices remain in a nonconductive state until the predetermined voltage is reached across the terminal, at this point the internal of the device will become a conductor. Excessive voltages and associated currents are shorted to ground.

2.6.2.10. *Back-up (Standby) Power Supplies*

They consist of a standby battery and an inverter which come into operation when the a.c. power supply is interrupted. Often such products incorporate clamping device on the line to attenuate spikes, surges and impulses. During normal operation of these devices, the load (equipment) is powered directly from the a.c.

2.7. UNINTERRUPTIBLE POWER SUPPLY (UPS)

The varied types of UPSs and their attributes often cause confusion in the data center industry. For example, it is widely believed that there are only two types of UPS systems, namely standby UPS and online UPS. These two commonly used terms do not correctly describe many of the UPS systems available. Many misunderstandings about UPS systems are cleared up when the different types of UPS topologies are properly identified. UPS topology indicates the basic nature of the UPS design. Various vendors routinely produce models with similar designs, or topologies, but with very different performance characteristics.

Common design approaches are reviewed here, including brief explanations about how each topology works. This will help you to properly identify and compare systems.

2.7.1. UPS Types

A variety of design approaches are used to implement UPS systems, each with distinct performance characteristics. The most common design approaches are as follows :

- Standby
- Line Interactive
- Standby-Ferro
- Double Conversion On-Line
- Delta Conversion On-Line

2.7.2. The Standby UPS

The Standby UPS is the most common type used for Personal Computers. In the block diagram illustrated in Fig. 11, the transfer switch is set to choose the filtered A.C. input as the primary power source (solid line path), and switches to the battery/inverter as the backup source should the primary source fail. When that happens, the transfer switch must operate to switch the load over to the battery/inverter backup power source (dashed path). The inverter only starts when the power fails, hence the name "Standby." High efficiency, small size, and low cost are the main benefits of this design. With proper filter and surge circuitry, these systems can also provide adequate noise filtration and surge suppression.

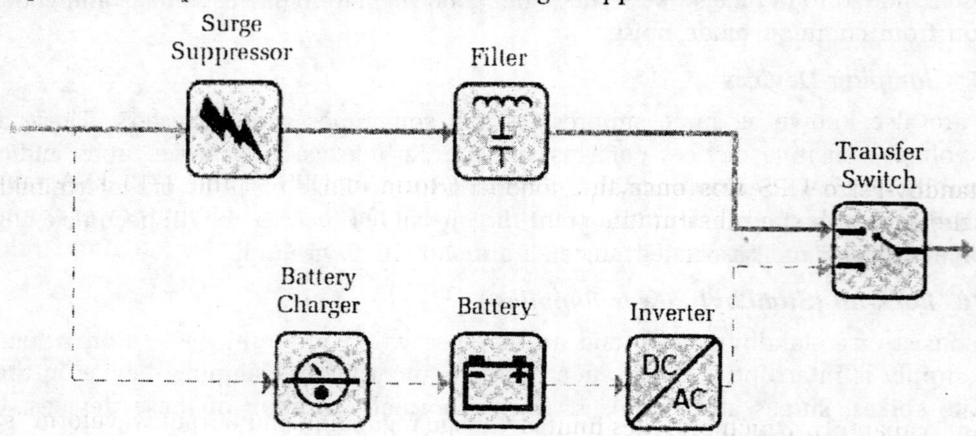

Fig. 11. *Standby UPS.*

2.7.3. The Line Interactive UPS

The Line Interactive UPS, illustrated in Fig. 12, is the most common design used for small business, Web, and departmental servers. In this design the battery-to-A.C. power converter (inverter) is always connected to the output of the UPS. Operating the inverter in reverse during times when the input A.C. power is normal provides battery charging.

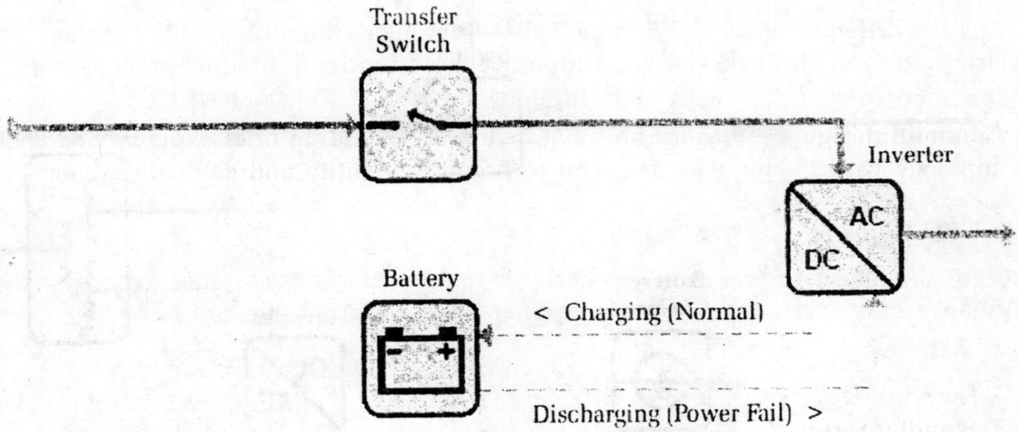

Fig. 12. *Line interactive UPS.*

When the input power fails the transfer switch opens and the power flows from the battery to the UPS output. With the inverter always on and connected to the output, this design provides additional filtering and yields reduced switching transients when compared with the Standby UPS topology.

In addition, the Line Interactive design usually incorporates a tap-changing transformer. This adds voltage regulation by adjusting transformer taps as the input voltage varies. Voltage regulation is an important feature when low voltage conditions exist, otherwise the UPS would transfer to battery and then eventually down the load. This more frequent battery usage can cause premature battery failure. However, the inverter can also be designed such that its failure will still permit power flow from the A.C. input to the output. which eliminates the potential of single point failure and effectively provides for two independent power paths. High efficiency, small size, low cost and high reliability coupled with the ability to correct low or high line voltage conditions make this the dominant type of UPS in the 0.5–5 kVA power range.

2.7.4. The Standby-Ferro UPS

The Standby-Ferro UPS was once the dominant form of UPS in the 3–15 kVA range. This design depends on a special saturating transformer that has three windings (power connections). The primary power path is from A.C. input, through a transfer switch, through the transformer, and to the output. In the case of a power failure, the transfer switch is opened, and the inverter picks up the output load.

In the Standby-Ferro design, the inverter is in the standby mode, and is energized when the input power fails and the transfer switch is opened. The transformer has a special "Ferro-resonant" capability, which provides limited voltage regulation and output waveform "shaping". The isolation from A.C. power transients provided by the Ferro transformer is as good as or better than any filter available. But the Ferro transformer itself creates severe output voltage distortion and transients, which can be worse than a poor A.C. connection. Even though it is a standby UPS by design, the Standby-Ferro generates a great deal of heat because the Ferro-resonant transformer is inherently inefficient. These transformers are also large relative to regular isolation transformers; so Standby-Ferro UPS are generally quite large and heavy.

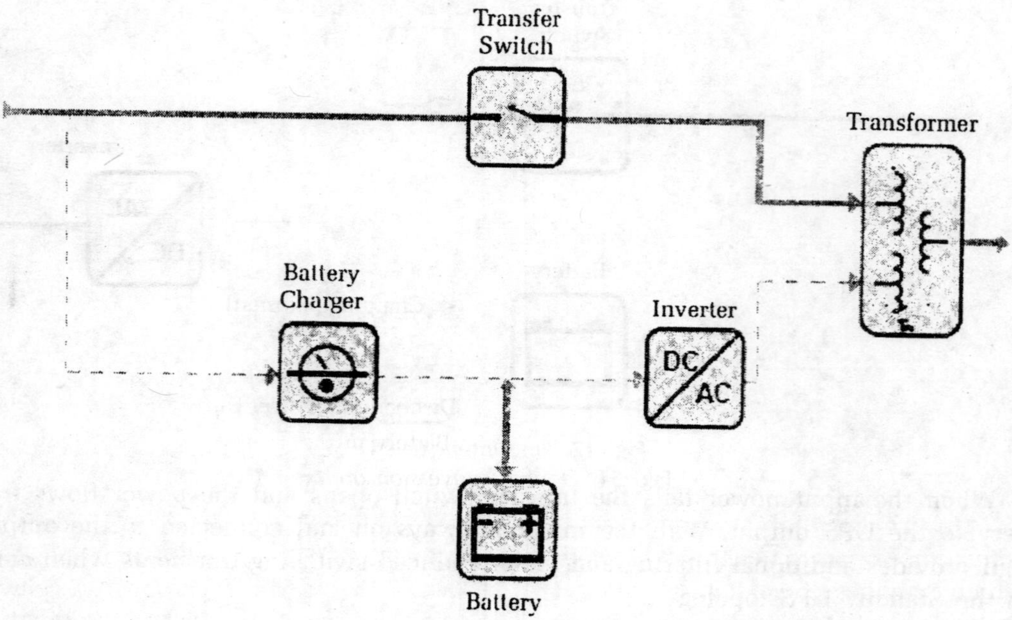

Fig. 13. *Standby-Ferro UPS.*

Standby-Ferro UPS systems are frequently represented as On-Line units, even though they have a transfer switch, the inverter operates in the standby mode, and they exhibit a transfer characteristic during an AC power failure. Fig. 13 illustrates this Standby-Ferro topology.

High reliability and excellent line filtering are this design's strengths. However, the design has very low efficiency combined with instability when used with some generators and newer power-factor corrected computers, causing the popularity of this design to decrease significantly.

The principal reason why Standby-Ferro UPS systems are no longer commonly used is that they can be fundamentally unstable when operating a modern computer power supply load. All large servers and routers use "Power Factor Corrected" power supplies which draw only sinusoidal current from the utility, much like an incandescent bulb. This smooth current draw is achieved using capacitors, devices which 'lead' the applied voltage, Ferro resonant UPS system utilize heavy core transformers which have an inductive characteristic, meaning that the current 'lags' the voltage. The combination of these two items form which is referred to as a 'tank' circuit. Resonance or 'ringing' in a tank circuit can cause high currents, which jeopardize the connected load.

2.7.5. The Double Conversion On-Line UPS

This is the most common type of UPS above 10 kVA. The block diagram of the Double Conversion On-Line UPS, illustrated in Fig. 14, is the same as the Standby, except that the primary power path is the inverter instead of the A.C. main.

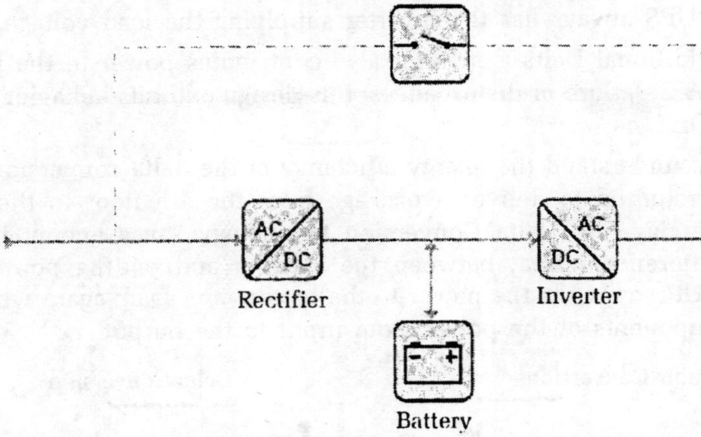

Fig. 14. *Double conversion on-line UPS.*

In the Double Conversion On-Line design, failure of the input A.C. does not cause activation of the transfer switch, because the input A.C. is charging the backup battery source which provides power to the output inverter. Therefore, during an input A.C. power failure, on-line operation results in no transfer time. Both the battery charger and the inverter convert the entire load power flow in this design, resulting in reduced efficiency with its associated increased heat generation.

This UPS provides nearly ideal electrical output performance. But the constant wear on the power components reduces reliability over other designs and the energy consumed by the electrical power inefficiency is a significant part of the life-cycle cost of the UPS. Also, the input power drawn by the large battery charger is often non-linear and can interfere with building power wiring or cause problems with standby generators.

2.7.6. The Delta Conversion On-Line UPS

This UPS design, illustrated in Fig. 15, is a newer, 10 year old technology introduced to eliminate the drawbacks of the Double Conversion On-Line design and is available in sizes

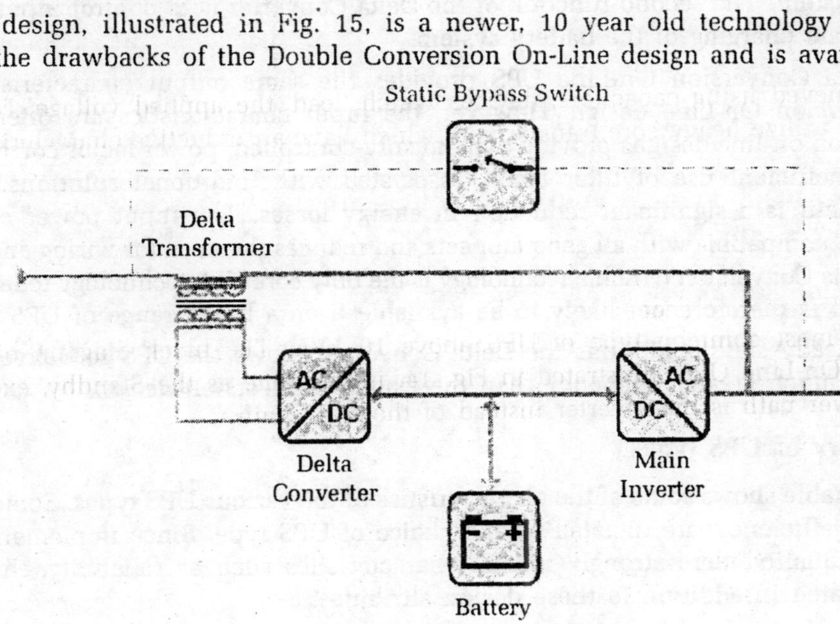

Fig. 15. *Delta conversion on-line UPS.*

ranging from 5 kVA to 1.6 MW. Similar to the Double Conversion On-Line design, the Delta Conversion On-Line UPS always has the inverter supplying the load voltage.

However, the additional Delta Converter also contributes power to the inverter output. Under conditions of A.C. failure or disturbances, this design exhibits behavior identical to the Double Conversion On-Line.

A simple way to understand the energy efficiency of the delta conversion topology is to consider the energy required to deliver a package from the 4th floor to the 5th floor of a building as shown in Fig. 16. Delta Conversion technology saves energy by carrying the package only the difference (delta) between the starting and ending points. The Double Conversion On-Line UPS converts the power to the battery and back again whereas the Delta Converter moves components of the power from input to the output.

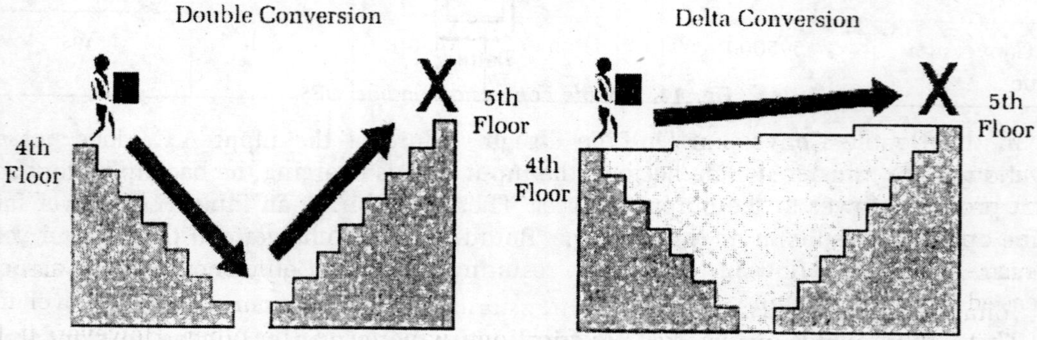

Fig. 16. *Analogy of double conversion versus delta conversion.*

In the Delta Conversion On-Line design, the Delta Converter acts with dual purposes. The first is to control the input power characteristics. This active front end draws power in a sinusoidal manner, minimizing harmonics reflected onto the utility. This ensures optimal utility and generator system compatibility, reducing heating and system wear in the power distribution system. The second function of the Delta Converter is to control input current in order to regulate charging of the battery system.

The Delta Conversion On-Line UPS provides the same output characteristics as the Double Conversion On-Line design. However, the input characteristics are often different. Delta conversion on-line designs provide dynamically-controlled, power factor corrected input, without the inefficient use of filter banks associated with traditional solutions. The most important benefit is a significant reduction in energy losses. The input power control also makes the UPS compatible with all generator sets and reduces the need for wiring and generator oversizing. Delta Conversion On-Line technology is the only core UPS technology today protected by patents and is therefore not likely to be available from a broad range of UPS suppliers.

During steady state conditions the Delta Converter allows the UPS to deliver power to the load with much greater efficiency than the Double Conversion design.

2.7.7. Summary of UPS types

The following table shows some of the characteristics of the various UPS types. Some attributes of a UPS, like efficiency, are dictated by the choice of UPS type. Since implementation and manufactured quality more strongly impact characteristics such as reliability, these factors must be evaluated in addition to these design attributes.

Table 2 : Comparison of Various UPS's

	Practical Power Range (kVA)	*Voltage Conditioning*	*Cost per VA*	*Efficiency*	*Inverter always operating*
Standby	0–0.5	Low	Low	Very High	No
Line Interactive	0.5–5	Design Dependent	Medium	Very High	Design Dependent
Standby Ferro	3–15	High	High	Low-Medium	No
Double Conversion On-Line	5–5000	High	Medium	Low-Medium	Yes
Delta Conversion On-Line	5–5000	High	Medium	High	Yes

2.8. ISOLATION TRANSFORMERS

Isolation transformers, as their name indicates, have primary and secondary windings, which are separated by an insulating or isolating medium. Isolation transformers do not help in curing voltage sags or swells; they merely transform the voltage from a primary level to a secondary level to enable power transfer from one winding to the other. However, if the problem is due to common mode noise, isolation transformers help to minimize noise coupling, and shielded isolation transformers can help to a greater degree. Common mode noise is equally present in the line and the neutral circuits with respect to ground. Common mode noise may be converted to transverse mode noise (noise between the line and the neutral) in electrical circuits, which is troublesome for sensitive data and signal circuits. Shielded isolation transformers can limit the amount of common mode noise converted to transverse mode noise. The effectiveness with which a transformer limits common mode noise is called *attenuation* (A) and is expressed in decibels (dB) :

$$A = 20 \log (V_1/V_2)$$

where V_1 is the common mode noise voltage at the transformer primary and V_2 is the differential mode noise at the transformer secondary. Fig. 17 shows how common mode noise attenuation is obtained by the use of a shielded isolation transformer. The presence of a shield between the primary and secondary windings reduces the inter-winding capacitance and thereby reduces noise coupling between the two windings.

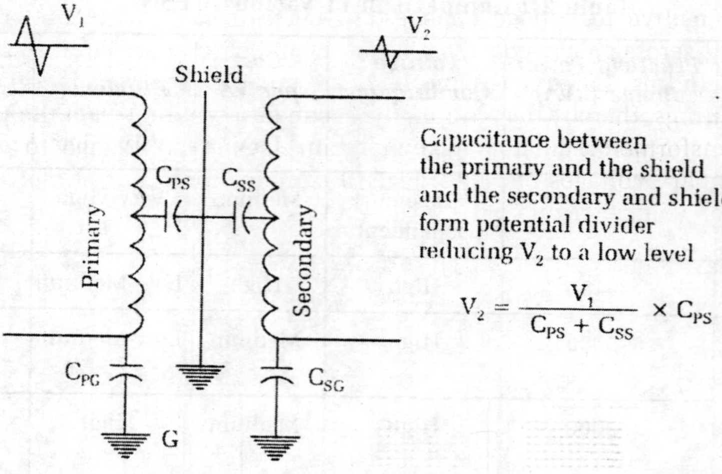

Fig. 17. *Common mode noise attenuation by shielded isolation transformer.*

2.9. VOLTAGE REGULATORS

Voltage regulators are devices that can maintain a constant voltage (within tolerance) for voltage changes of predetermined limits above and below the nominal value. A switching voltage regulator maintains constant output voltage by switching the taps of an autotransformer in response to changes in the system voltage, as shown in Fig. 18.

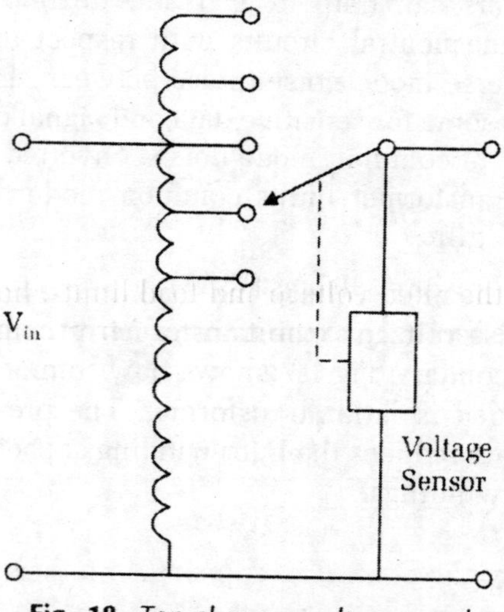

Fig. 18. *Tap-changer voltage regulator.*

The electronic switch responds to a signal from the voltage-sensing circuitry and switches to the tap connection necessary to maintain the output voltage constant. The switching is typically accomplished within half of a cycle, which is within the ride-through capability of most sensitive devices.

Ferro-resonant voltage regulators are static devices that have no moving components. They operate on the principle that, in a transformer, when the secondary magnetic circuit is operating in the saturation region the secondary winding is decoupled from the primary and therefore is not sensitive to voltage changes in the primary. The secondary winding has a capacitor connected across its terminals that forms a parallel resonant circuit with the inductance of the secondary winding. Large magnetic fields are created in the magnetic core surrounding the secondary windings, thereby decoupling the secondary winding from the primary. Typically Ferro-resonant transformer regulators can maintain secondary voltage to within ±0.5% for changes in the primary voltages of ±20%. Fig. 19 contains the schematic of a Ferro-resonance transformer type of voltage regulator.

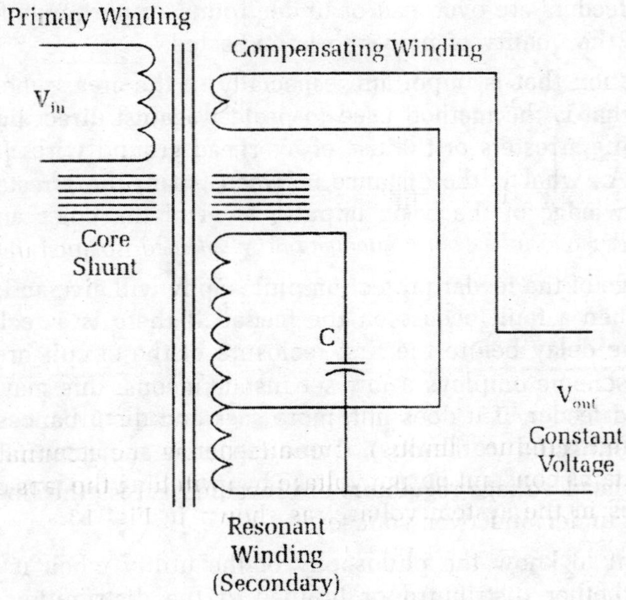

Fig. 19. *Ferro-resonant transformer.*

Ferro-resonance transformers are sensitive to loads above their rated current. In extreme cases of overload, secondary windings can become detuned, at which point the output of the transformer becomes very low. Voltage sags far below the rated level can also have a detuning effect on the transformer. Within the rated voltage and load limits, however, the Ferro-resonance transformer regulators are very effective in maintaining fairly constant voltage levels.

2.10. INDICATORS OF THE QUALITY OF POWER IN A PLANT

A few questions asked by the user of power to the utility company, which can give an indication of the quality of power expected to appear at the service entrance board. These questions are :

1. The length of the feeder to which your electric power supply is connected to and the point of supply in conjunction (in relation) to the total length of the feeder. Example : 12 miles, 24 miles, 6 miles etc. In theory the longer the feeder the more exposed it will be and thus more faults will be seen due to trees, animals, accidents, failure of components etc. Also, this will give an indication of potential under-voltage or over-voltage conditions under emergency overload conditions or under-load conditions.
2. What is the philosophy of the utility regarding the use of reclosers for longer feeders ? For example, if a feeder is 24 miles in length and the supply point is at 8 miles from

the transformer or distribution station, if there is a reclosers midway on this feeder and a fault occurs at 14 miles from the station, the reclosers will interrupt the fault rather than the breaker at the station thus the outage will be only confined to the loads downstream the reclosers and the upstream loads would only see a brief sag.

3. What is the total number of feeders connected to the bus feeding the subject plant ? If a fault occurs on any feeder, all the loads connected to the faulted feeder and the other feeders connected to the bus will see a voltage sag or dip. A knowledge of whether the feeders are overhead or underground or portions of would, also, give an indication of the quality of power to be expected.
4. Another question that is important, especially in the areas where thunderstorms are frequent is what is the method used to protect against direct lightning hit ? Is it the use of lightning arresters or the use of overhead ground wires (for shielding). If it is the use of L.A., what is the distance between (span) the arresters used. For indirect hits, the knowledge of the basic impulse level of the poles and insulators used is usually useful.
5. The knowledge of the feeder protection philosophy will give an indication of duration of outages when a fault occurs on the feeder. If there is a reclosing scheme in the protection, the delay before the first reclosure of the circuit breaker is important to know. If the scheme employs a low set instantaneous, this may cause more outages to the affected feeder. If it does not, more sags and disturbances will be imposed on all feeders (connected to the bus), if one feeder is subjected to a fault.
6. Whether overhead voltage regulators are installed across the lines, as this may affect conditions of under and over voltages.
7. It is important to know the philosophy of the utility when it comes to the use of capacitors (whether distributrd or lumped at the distribution station, transformer station or at the transmission level). As this may affect the spikes, surges or impulses that the connected equipment, at the lower level, may see. The magnification at the lower voltage levels will be severe under the following conditions : the capacitor switched on the higher voltage system has a rating (KVAR) much higher than the capacitor at the low voltage thus, the frequency of oscillation which occurs when the high voltage capacitor is energized is close to the resonant frequency formed by the step down transformer in series with the low voltage capacitor and if there is little damping presented by the loads on the low voltage system (it is usual with industrial plants as induction motors do not provide significant damping to such transients).

SUMMARY

- The disturbances are introduced into the power distribution systems either from the load centre or the utility including environmental causes.
- Voltage Swell is a rapid short-term increase in voltage.
- Voltage Sag is a rapid short-term decrease in voltage.
- Transient is an extremely high and nearly instantaneous increase in voltage.
- Noise a disturbance in the smooth flow of electricity.
- Interruption is the total loss of power for some period of time.

- Phase controlled devices will vary the relative magnitudes of the harmonic components.
- The interface between the mains network and any electronic equipment is a switch mode power supply (SMPS).
- When transformers are energized, inrush currents that are rich in second harmonics will be seen by the primary of the transformer.
- The major effect of line disturbances on these machines is increased heating.
- Voltage harmonics cause higher transformer voltage and insulation stress.
- Power electronic equipment is susceptible to misoperation caused mostly by voltage notching.
- Capacitors with the inductance of the system can cause resonance.
- Series filters consist of a parallel LC circuit.
- Harmonics currents can be eliminated by inducing harmonic flux.
- Voltage regulators are devices that can maintain a constant voltage (within tolerance) for voltage changes of predetermined limits above and below the nominal value.

Review Questions and Exercises

1. Explain common power frequency disturbances. *(MDU, Rohtak, May 2008)*
2. Describe the operation of a voltage regulator with diagram. *(MDU, Rohtak, May 2008)*
3. Explain static uninterruptible power source system. *(MDU, Rohtak, May 2008, Dec. 2008, May 2009)*
4. Describe the function and properties of an Isolation transformer. *(MDU, Rohtak, May 2008)*
5. Write a short note on voltage sag and voltage swell.
6. Describe briefly about SMPS.
7. What are the techniques used to reduce disturbance to improve power quality ?
8. Explain various regulators. *(MDU, Rohtak, Dec. 2008)*
9. Write short note on Isolation Transformers. *(MDU, Rohtak, Dec. 2008)*

□□□

Chapter 3 ELECTRICAL TRANSIENTS

3.1. INTRODUCTION

A transient occurs in the power system when the network changes from one steady state into another. This can be, for instance, the case when lightning hits the ground in the vicinity of a high-voltage transmission line or when lightning strikes a substation directly. The majority of power system transients are, however, the result of a switching action. Loadbreak switches and disconnectors switch off and switch on the parts of the network under load and no-load conditions. Fuses and circuit breakers interrupt higher currents and clear short-circuit currents flowing in faulted parts of the system. The time period when transient voltage and current oscillations occur is in the range of microseconds to milliseconds. On this timescale, the presence of a short-circuit current during a system fault can be regarded as a steady-state situation, wherein the energy is mainly in the magnetic field, and when the fault current has been interrupted the system is transferred into another steady-state situation, wherein the energy is predominantly in the electric field. The energy exchange from the magnetic field to the electric field is, when the system is visualised by lumped elements, noticed by transient current and voltage oscillations.

3.1.1. The Origins of Transient Overvoltages

Transient overvoltages in power systems originate from one cause, energy being injected into the power system, but from two sources : lightning discharges, or switching within the power system. However, in communication and data systems, there is another source of transients : the coupling of power system transients into the data system.

Lightning discharges may not necessarily mean direct termination of a lightning stroke onto the power system. A lightning stroke terminating near a power line, either by hitting a tree or the bare earth, will create a very fast-changing magnetic field that can induce voltages—and inject energy—into the loop formed by the conductors of the power system. Lightning can also inject overvoltages in a power system by raising the ground potential on the surface of the earth where the stroke terminates. while more distant "ground" points remain at a lower voltage, closer to the potential of "true earth." The literature and other presentations in this conference provide information on the characteristics of lightning discharges.

Surges from power system switching create overvoltages as a result of either trapped energy in loads being switched off or restrikes in the switchgear.

3.2. TYPES AND CAUSES OF TRANSIENTS

3.2.1. Impulsive Transient

An impulsive transient is a brief, unidirectional variation in voltage, current, or both on a power line. The most common sources of impulsive transients are lightning strikes (Fig. 1). Impulsive transients due to lightning strikes can occur because of a direct strike to a power line or from magnetic induction or capacitive coupling from strikes on adjacent lines. Note that in this case, the maximum amplitude of the transient current is ~23 kiloamps and the duration of the transient current is tens of microseconds. The frequency and amplitude of lightning-induced transients vary geographically, as shown in Fig. 2.

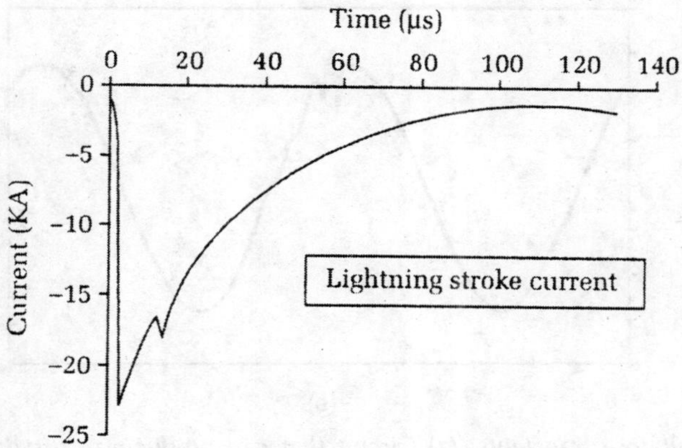

Fig. 1. *Transient due to lightning strike.*

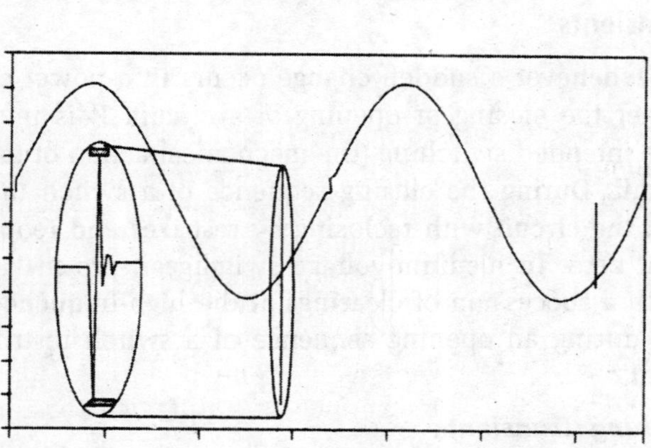

Fig. 2. *Impulsive transient.*

3.2.2. Oscillatory Transient

An oscillatory transient is a brief, bidirectional variation in voltage, current, or both on a power line. These transients can occur due to resonances during switching. A circuit capable of exhibiting this phenomenon is shown in Fig. 3(*a*). A power supply bus is shown with the bus having inductance *L*. A capacitor bank labeled C_1 is connected at one end of the bus. This capacitor bank may be in place, for instance, for power factor improvement or for voltage sag improvement. If at some time, we switch in capacitor bank C_2 with the switch as shown, a resonant condition is set up between the line inductance and the capacitor banks. The resultant resonance will be underdamped, and the current in the capacitor bank may look something like that in Fig. 3(*b*).

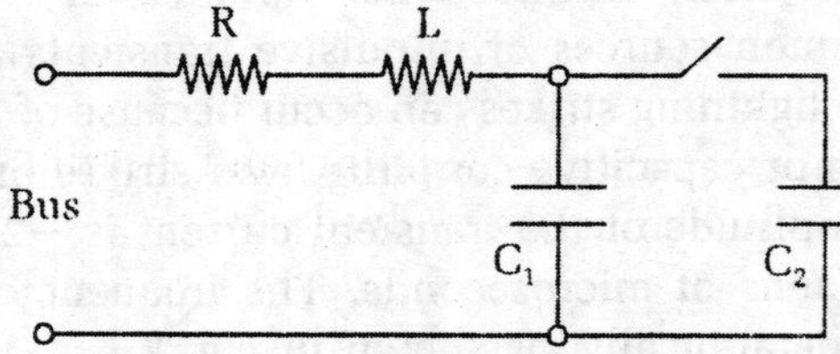

(*a*)

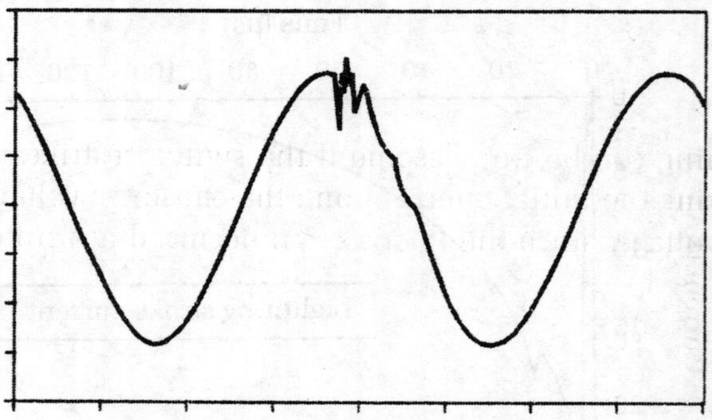

(*b*)

Fig. 3. *Oscillatory transients. (a) Circuit that can produce an oscillatory transient. (b) An oscillatory transient.*

3.2.3. Switching Transients

A transient is created whenever a sudden change occurs in a power circuit, especially during power switching—either the closing or opening of a circuit. It is important to recognize the difference between the intended switching (the mechanical action of the switch) and the actual happening in the circuit. During the closing sequence of a switch the contacts may bounce, producing openings of the circuit with reclosing by restrikes and reopening by clearing at the high-frequency current zero. In medium-voltage switchgear, pre-strikes can occur just before the contacts close, with a succession of clearings at the high-frequency current zero, followed by restrike. Similarly, during an opening sequence of a switch restrikes can cause electrical closing(s) of the circuit.

3.2.3.1. Simple Switching Transients

Simple switching transients include circuit closing transients. Transients initiated by clearing a short circuit and transients produced when the two circuits on either side of the switch being opened oscillate at different frequencies. In circuits having inductance and capacitance (all physical circuits have at least some in the form of stray capacitance and inductance) with little damping, these simple switching transients are inherently limited to twice the peak amplitude of the steady-state sinusoidal voltage. Another limit to remember when analyzing transients associated with current interruption (circuit opening) is that the circuit inductance tends to maintain the current constant. At most, then, a surge protective device provided to divert the current will be exposed to that initial current. Without a surge protective device the current is available to charge the circuit capacitances at whatever voltage is required to store the inductive energy from the current by converting it into capacitive energy.

3.2.3.2. Abnormal Switching Transients

Several mechanisms are encountered in practical power circuits. These mechanisms can produce transient overvoltages far in excess of the theoretical twice the normal limit. Two such mechanisms occur frequently : current chopping and restrikes, the latter being especially troublesome when capacitor switching is involved.

Current chopping is the name given to the rapid current reduction which fuses or which circuit breakers can force when clearing a circuit prior to the natural current zero of the power system. When there is inductance in the circuit, this rapid current change can produce high overvoltages—some 10 times the normal circuit voltage. A classical example is an unloaded transformer where the magnetizing inductance is high and the energy stored in the magnetic core can charge only the winding capacitance.

Capacitor switching can be troublesome if the switch restrikes after current interruption. Restrike occurs following the initial interruption : the capacitor voltage remains nearly constant at maximum system voltage, since the interruption occurred at zero current which is 90° apart from the voltage zero, while the system voltage follows the normal sine wave. At 180° after interruption, the switch has to support twice the amount of system voltage, a stress it might not be able to withstand with its contacts not completely separated. A restrike can occur under these circumstances. In such a case, the capacitance of the circuit will tend to drive the voltage not toward the system voltage but beyond it—theoretically up to twice the difference. Such an overshoot means a possible voltage of three times the system voltage. While this high-frequency oscillation takes place the switch may clear at a high-frequency current zero only to restrike again later with an even greater difference of voltage and then escalating to an even higher overshoot. The outcome will be either a breakdown in one of the components or it the switch eventually recovers enough dielectric withstand in its opening gap, no further restrikes. But severe over-voltages will have been impressed on the system.

A similar scenario can unfold when an ungrounded power system experiences an arcing ground fault. The switching action is then not the result of a deliberate parting of contacts but the intermittent connection produced by the arc.

These switching overvoltages high as they may be, are somewhat predictable and can be estimated with reasonable accuracy from the circuit parameters, once the mechanism involved has been identified. There is still some uncertainty as to where and when they occur because the worst offenders result from some abnormal behavior of a circuit element. Lightning-induced transients are much less predictable because there is a wide range of coupling possibilities. Moreover, one user, assuming that his system will not be the target of a direct hit may take a casual view of protection. While another, fearing his system will experience a "Worst case", may demand the utmost protection.

In response to these concerns various committees and working groups have attempted to describe ranges of transient occurrences or maximum values occurring in power circuits.

3.2.4. Capacitor Switching Transients

Transient disturbances in power systems may damage key equipment, potentially having a great impact on system reliability. These transients may be introduced during normal switching operations, interruption of short circuits, lightning strikes, or due to equipment failure. Phasor analysis or other simplified analysis methods are usually inadequate due to system frequency dependencies and nonlinearities. Therefore, time-domain computer models are typically developed as a means of predicting the severity of the transient occurrences. The simulations are typically performed using simulation software such as the Electromagnetic Transient Program (EMTP). In this work, a royalty-free version of EMTP, called the Alternative Transients Program (ATP) was used.

During the switching of shunt capacitor banks, high magnitude and high frequency transients can occur. In earlier years, shunt capacitor banks have been more commonly installed at distribution and lower sub-transmission levels. However, there has been a recent proliferation of new capacitor banks at transmission levels. Since larger higher-voltage capacitor banks have more stored energy and there is a higher system X/R ratio and less damping.

3.2.4.1. Conceptual Introduction

The 34.5-kV per-phase system of Fig. 4 is used to provide a conceptual introduction to some of the common transients involved in capacitor bank switching. R_1 and L_1 represent the system source impedance. CB4 feeds two capacitor banks, represented by C_1 and C_2. S_1 and S_2 represent the circuit breakers used to switch the capacitor banks. LB is the inductance of the bus spanning between the capacitor banks. R_2 and L_2 are the total impedance of the feeder and distribution transformer. A distribution-level capacitor bank is attached to the transformer secondary. CB3 can be used to initiate and interrupt a ground fault on the bus at some distance down the feeder, depending on location of the ground. Using different portions of this system, five transients can be addressed : 1) energization inrush, 2) back-to-back energization, 3) outrush into a nearby fault, 4) voltage magnification, and 5) transient recovery voltage (TRV).

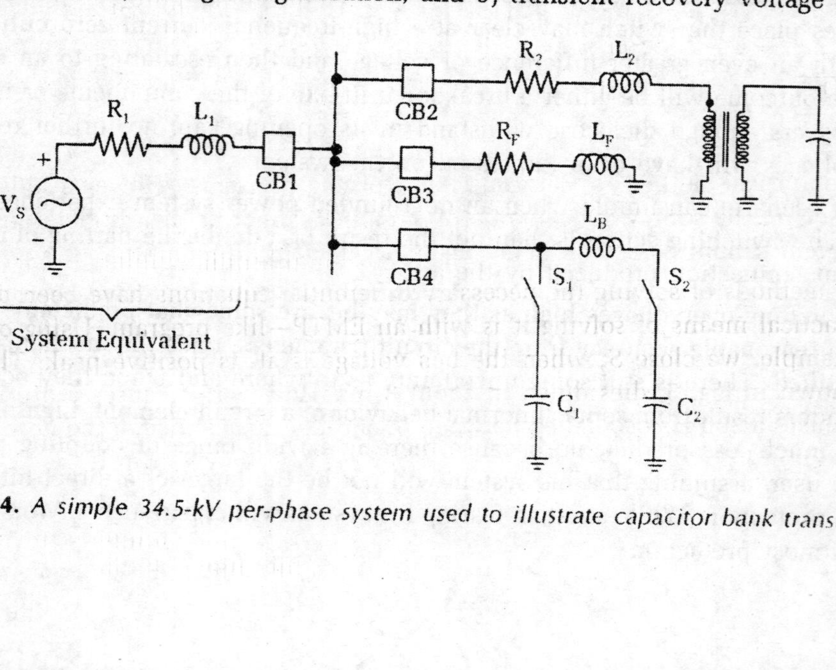

Fig. 4. *A simple 34.5-kV per-phase system used to illustrate capacitor bank transients.*

3.2.4.2. *Energization Inrush*

Energization inrush is a transient occurring when the first (or only) bank at the bus is energized. The transient is characterized by a surge of current having a high magnitude and a frequency as high as several hundred Hertz. There is also a transient overvoltage on the bus, caused by the surge of inrush current coming from the system source. To illustrate, we begin with *CB*1 and *CB*4 closed, energizing C_1 by closing S_1. The operative circuit consists of a Thevenin source in series with C_1. Analysis of switching transients typically makes use of superposition, breaking the transient into a) natural response caused by the switching and b) the forced response due to the 50-Hz system source. If we neglect system resistance, the natural response component of the inrush current into the capacitor may be approximated as :

$$i(t) = \frac{V(0)}{Z_0}\sin\omega_0 t,$$

where $Z_0 = \sqrt{\dfrac{L}{C_1}}, \quad \omega_0 = \dfrac{1}{\sqrt{LC_1}}.$

And $V(0)$ is difference between the source voltage and the initial voltage of the capacitor at the instant of energization. It can be seen that as a capacitor bank size increases, peak inrush current increases and its frequency decreases.

3.2.4.3. *Back-to-Back Switching*

Energizing the second bank C_2 when the first bank C_1 is already energized is called back-to-back switching, and is simulated by closing switch S_2 when C_1 is already operating in steady state. The resulting inrush to C_2 is a high-frequency transient which primarily involves the series combination of C_1, *LB*, and C_2, driven by the voltage $V(0)$ on C_1 at the instant S_2 is closed. Just as with energization of a single bank, worst case occurs when the bus voltage is at its positive or negative peak value. This transient is typically in the kHz range. The expression for the high-frequency transient current riding on the steadystate 50-Hz response is :

$$i(t) = \frac{V(0)}{Z_{01}}\sin\omega_{01} t$$

where, $Z_{01} = \sqrt{\dfrac{L_B}{C}}, \quad \omega_{01} = \dfrac{1}{\sqrt{L_B C_{EQ}}}$ and $C_{EQ} = \dfrac{C_1 C_2}{C_1 + C_2}.$

There are empirical methods to calculate the expected transients. However, this approach neglects the effects of the source and the rate of damping cannot be clearly determined. Such a high-order system cannot be solved using manual calculations.

Numerical methods of solving the necessary differential equations have been developed, but the only practical means of solving it is with an EMTP—like program. Using our simple circuit as an example, we close S_2 when the bus voltage is at its positive peak. The current inrush to C_2, shown in Fig. 5, dies down in about 3 ms. However, it causes a dip in the bus voltage which causes a 300-Hz oscillation between L_1 and the capacitance of the now-paralleled banks $C_1 + C_2$ (Fig. 6).

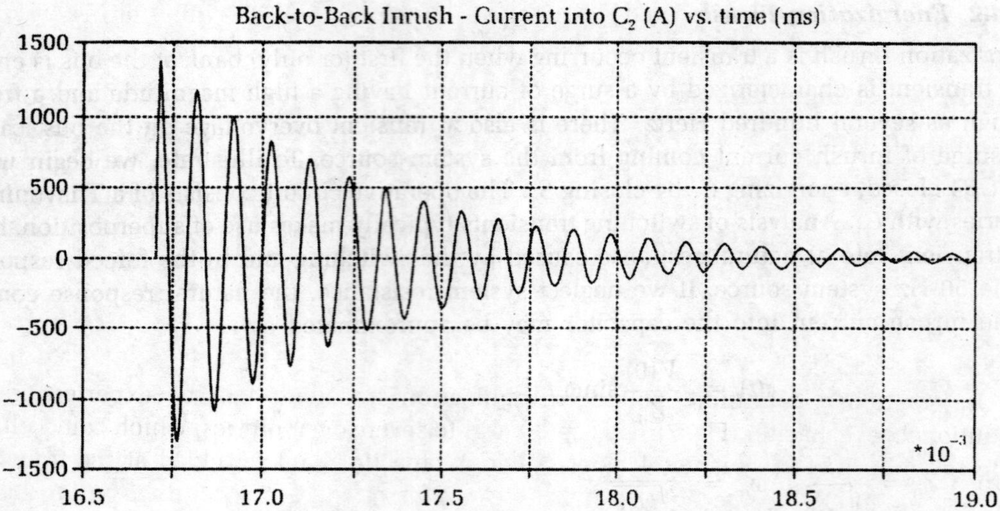

Fig. 5. *Back-to-back energization. Current inrush into C_2.*

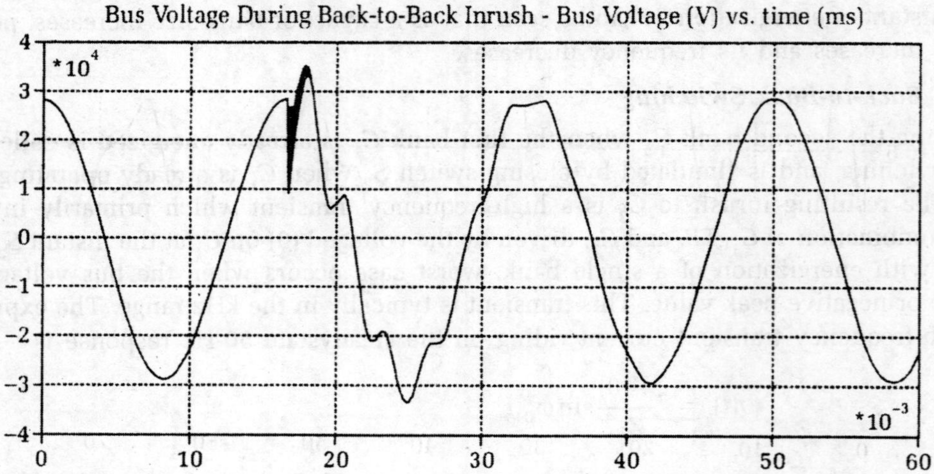

Fig. 6. *Transient in bus voltage following back-to-back switching.*

3.2.4.4. Outrush Transient

With capacitor bank C_1 operating in steady state, *CB*3 can be closed, simulating a fault at some distance down the local feeder. C_1 discharges into the fault, resulting in a damped oscillation with LF. The outrush current from the capacitor is given by :

$$i(t) = \frac{V(0)}{Z_{02}} \sin \omega_{02} t$$

where, $Z_{02} = \sqrt{\frac{L_F}{C_1}}$, $\omega_{02} = \frac{1}{\sqrt{L_F C_1}}$, and

$V(0)$ is the instantaneous magnitude of the voltage across capacitor C_1 at the instant of the fault. As the fault location is moved back toward the bus, *LF* decreases, increasing the current magnitude and frequency of the outrush current. If a bus fault occurs, inductance is on the order of tens of micro-Henries, and outrush can be severe. Outrush-limiting reactors can be installed in series with each capacitor bank if necessary.

3.2.4.5. Voltage Magnification

This transient manifests itself as a voltage increase when a capacitor bank is energized. A common scenario is the interaction between a distribution-level capacitor bank and another nearby bank on the transmission system. To reproduce it with our simple circuit of Fig. 4, we close *CB2* with C_1 already energized. If the natural frequency of L_1 and C_1 matches the natural frequency of L_2 and the distribution capacitor bank *CLV*, then voltage magnification may occur.

$$\omega_0 = \frac{1}{\sqrt{L_2 C_{LV}}} = \frac{1}{\sqrt{L_1 C_1}}$$

Under such a condition, voltage magnification transients can be experienced at the distribution level capacitor (Fig. 7). This could lead to severe overvoltages, which could ultimately lead to the failure of the capacitor bank. When a capacitor bank is added at the transmission level, it is important to rule out or mitigate any interaction with local distribution banks.

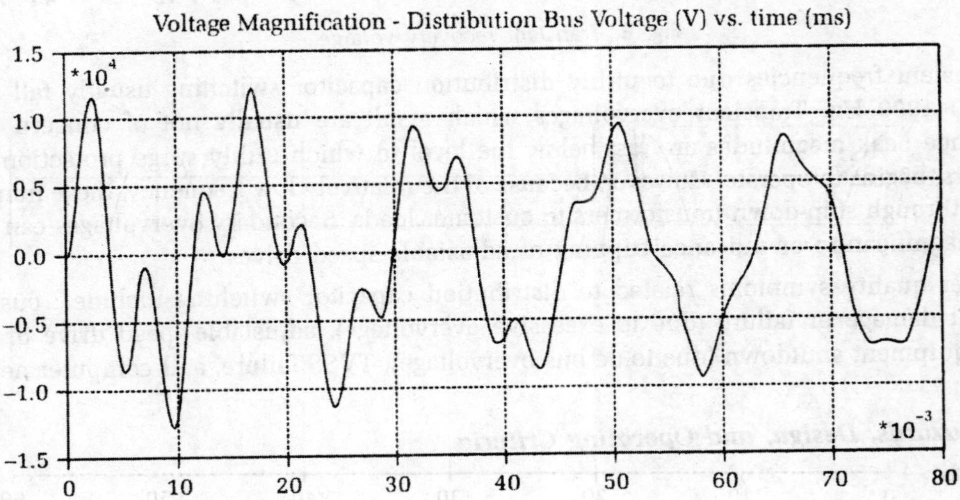

Fig. 7. *Voltage magnification at distribution bus.*

3.2.4.6. Transient Recovery Voltage (TRV)

This is the transient voltage that appears across the poles of a circuit breaker upon interrupting a circuit. This can be simulated in Fig. 4 by considering the voltage across the contacts of CB1 after clearing a fault on the feeder or on the bus. During the fault, the voltage at the bus is very low or zero. When the breaker contacts open, the voltage on the source side "recovers" or jumps back up to rated system voltage.

However, the voltage on the source side of the circuit breaker recovers by charging up the bushing capacitance of the circuit breaker. Additional shunt capacitances may also be present in the form of stray capacitance or capacitor banks on the source side of the circuit breaker. The total shunt capacitance present, as it is charged, resonates with the source inductance L_1.

If there is only bushing capacitance, the natural frequency is very high. If there is little damping (*i.e.*, if R_1 is small), the peak voltage across the circuit breaker can reach 2.0 pu. The results of a simulation using our simple system are shown in Fig. 8. In this case, peak recovery voltage reaches 1.4 pu, and the frequency is over 5 kHz.

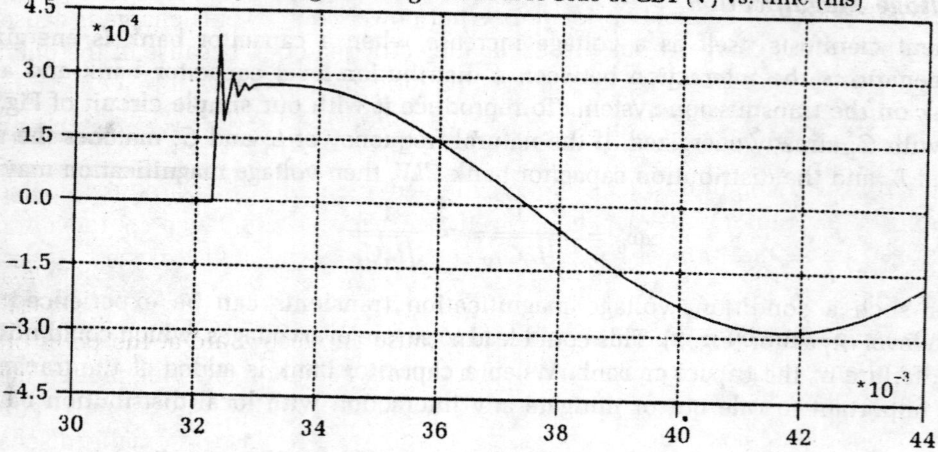

Fig. 8. *Transient recovery voltage.*

Transient frequencies due to utility distribution capacitor switching usually fall in the range 300–1000 Hz. Transient overvoltages which result are usually not of concern to the utility, since peak magnitudes are just below the level in which utility surge protection, such as arresters, begins to operate. However, because of the relatively low frequency, these transients will pass through step-down transformers to customer loads. Secondary overvoltages can cause voltage magnification or nuisance tripping of adjustable-speed drives.

Power quality symptoms related to distribution capacitor switching include : customer equipment damage or failure (due to excessive overvoltage), adjustable-speed drive or other process equipment shutdown (due to dc bus overvoltage), TVSS failure, and computer network problems.

3.2.4.7. *Hazards, Design, and Operating Criteria*

(*a*) General

Capacitor banks shall be isolated by elevation, NEC barriers, or enclosures to preclude accidental contact with charged terminals, conductors, or support structures. Enclosures and barriers shall be used to protect personnel from projectiles that might be expelled from the capacitors during a fault : for capacitor banks capable of storing more than 50 kJ of energy, special enclosure requirements may be required to provide protection.

Oil filled capacitors should be avoided in radiation environments. Experience has shown that the radiation breaks down the oil and causes a potential over pressure hazard. Access to capacitor areas shall be restricted until all capacitors have been discharged, shorted, and grounded.

All capacitors in storage shall be short-circuited with a conductor securely fastened to the terminals and left in place until the capacitors are used again or scrapped. Ventilation to keep the temperature of ambient air at capacitor installations at recommended levels shall be provided. Capacitor cases, unless obviously connected to a recognized grounding conductor or grounded structure, shall be considered "charged", and shall be grounded in the same manner as capacitor terminals.

(b) Stored Energy

Capacitors or capacitor banks with stored energy of 10 J or more constitute a lethal shock hazard. Although they have been disconnected and discharged, capacitors may accumulate a charge without benefit of connection to an external power source. This charge is caused by the slow release of electric charges from within the dielectric material, because of the phenomenon known as "dielectric absorption". It also is possible for capacitors to acquire a charge from local atmospheric electrical disturbances and by corona from a nearby high-voltage terminal, such as on an adjacent capacitor.

During transient conditions, capacitors could acquire a charge hazardous to personnel, or an over-voltage harmful to the capacitor itself because of inductance in the circuit. This inductance can be in the form of coils or magnets, or in wiring and leakage.

(c) Discharging

Discharging a capacitor by a grounding hook can cause an electric arc at the point of contact. Such release of energy can also cause burns from thermal radiation or flying molten metal. Any residual charge shall be removed from capacitors by grounding the terminals with a low-impedance grounding hook before beginning to work with them. Automatic discharge and grounding devices shall not be relied upon for personnel safety; grounding hooks must be used to ensure safe operations. Grounding hooks shall be inspected prior to use to ensure that all connections are secure, and that the grounding conductor is in good condition. Grounding the output typically will not discharge internal capacitor banks.

Short circuit all capacitors in storage with a conductor not smaller than 14 AWG, securely fastened to the terminals and left in place until the capacitors are to be used again.

(d) Connecting

A dangerously high voltage can exist across the impedance of a few feet of grounding cable at the moment of contact with a charged capacitor. Operating personnel shall stand clear of cables attached to grounding hooks at the moment of application to a capacitor terminal.

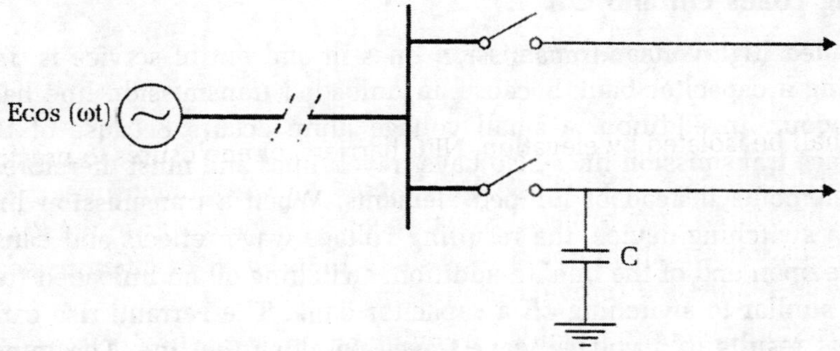

Fig. 9. *Single-line diagram for a substation with a capacitor bank for voltage regulation.*

(e) Safety Devices

Safety devices, such as shorting switches and grounding switches, and their associated cables and cable connectors, shall be designed to withstand the mechanical forces from the large currents which result from their operation.

Protective devices, such as automatic shorting switches and grounding hooks, shall be tested after installation, and at a minimum of every 3 years thereafter to verify their operation.

(*f*) Faults

Internal faults may rupture capacitor containers, particularly when many capacitors are connected in parallel. This rupture is normally caused by the boiling of the insulating liquid in the capacitor and may even occur where the peak fault current is not high. Metal case capacitors will usually swell and vent before large amounts of overpressure occur. Cast or phenolic cased units present a more serious hazard : the force of the explosion may cause serious injury.

Capacitors should be provided with current-limiting devices, such as fuses and resistors, which are capable of interrupting available fault current or limiting it to safe and manageable values. When this is not possible, alternate means to ensure personnel protection should be incorporated (*i.e.*, enclosure). Rupture of a container by an internal fault can create a fire hazard because combustible dielectric could be ignited.

(*g*) Fuses

Fuses may be used to interrupt the discharge of energy from a power source or a capacitor bank into a faulted individual capacitor. If fuses and the capacitors are not adequate for this application, they could explode, expelling dangerous projectiles. Fuses designed for AC operation depend on the current passing through zero on the next reversal of the line voltage to guarantee that the fuse will clear. Even if the correct DC-rated fuses are used, a complete capacitor bank may discharge through a fault at very high currents before the fuse clears. The fuse in this type of application usually does not allow a shorted capacitor to permanently load a power source that is feeding it, such as in factor correcting service on power transmission lines.

(*h*) Bleeder Resistors

It is essential that bleeder resistors be used on each capacitor that is fused to ensure that the capacitor discharges when it becomes isolated. If bleeder resistors are not used to discharge the capacitors, the capacitors must be automatically discharged. The residual voltage on capacitors must be reduced to 50 V or less, within one minute.

3.2.5. Switching Loads On and Off

Switching unloaded high-voltage transmission lines in and out of service is, in principle, the same as switching a capacitor bank because an unloaded transmission line has a dominantly capacitive behaviour. In addition, a small voltage jump occurs because of the Ferranti-rise effect. High-voltage transmission lines also have travel times and must therefore be represented by distributed elements instead of lumped elements. When a transmission line is energized after closing of a switching device, the resulting voltage wave reflects and causes doubling of the voltage at the open end of the line. In addition, switching off an unloaded transmission line is, in principle, similar to switching off a capacitor bank. The Ferranti rise causes a transient voltage jump that results in a voltage wave travelling along the line. The interrupting device is stressed by a voltage jump at the supply side and an oscillation at the line side. The resulting steep voltage jump across the contacts of the interrupting device can cause, because of the short arcing time, a re-ignition and prolong the arcing time. When we analyze three-phase switching of unloaded high-voltage transmission lines, the capacitive coupling between the phases and the capacitance to ground have to be taken into account (see Fig. 10).

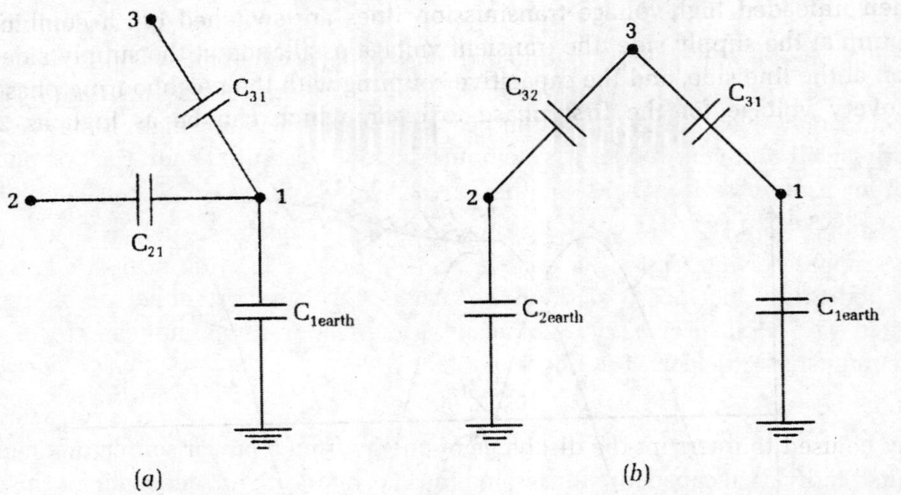

Fig. 10. *The coupling effect of the line capacitances when switching unloaded high-voltage transmission lines (a) phase 1 has been switched off and phases 2 and 3 are still at voltage (b) phases 1 and 2 are switched off and phase 3 is still at voltage.*

When the first phase (*e.g.*, phase 1) has cleared the power-system, voltage *Vi* of the neighbouring phases 2 and 3 is coupled into the line-side DC voltage V_1 with a coupling factor $k = Vi/V_1$. The coupling factor between the neighbouring phases and phase 1 depends on the ratio of C_{21}/C_1 earth and C_{31}/C_1 earth. When the second phase clears (let us assume that this is phase 2), then phase 3 is still at system voltage and this voltage couples into the line-side DC voltages of phase 1 and phase 2. The value of the coupling factor depends on the tower structure and circuit design but usually has a value between 0.2 and 0.4. For a double circuit, when two circuits are hanging in the same tower, the coupling factor can be higher when the neighbouring circuit is in operation. Capacitive coupling increases the TRV across the contacts of the interrupting device with the coupling factor (Fig. 11).

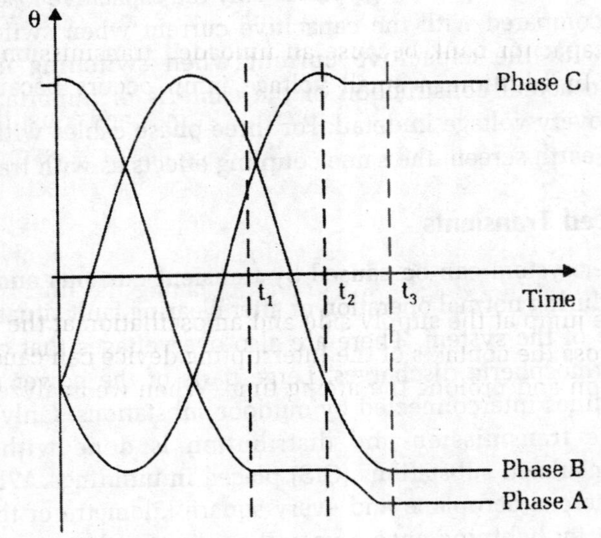

Fig. 11. *Line-side voltages after switching off an unloaded high-voltage transmission line. The three phases interrupt the capacitive current at t_1, t_2, and t_3.*

When unloaded high-voltage transmission lines are switched off, a combination of the voltage jump at the supply side, the transient voltage oscillation at the supply side, the voltage oscillation at the line side, and the capacitive coupling with the neighbouring phases can result in a recovery voltage for the first phase to clear, which can be as high as 2.8 per unit (Fig. 12).

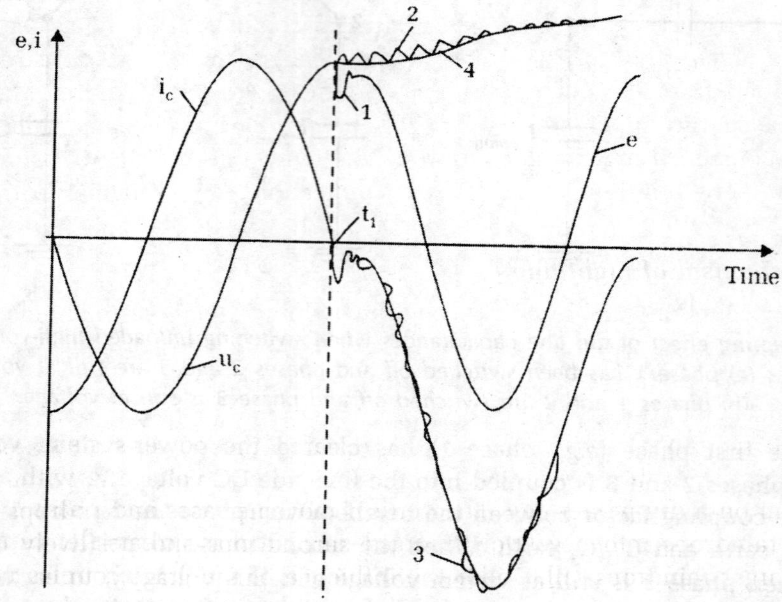

Fig. 12. *Current and voltage traces after switching off an unloaded high-voltage transmission line : 1 : voltage jump at the supply side; 2 : transient voltage oscillation at the supply side; 3 : voltage across the contacts of the interrupting device; 4 : power frequency voltage coupled in from the neighboring phase*

When switching unloaded high-voltage cables only the capacitive load current is interrupted. This current is small compared with the capacitive current when switching capacitor banks, but large compared with the capacitive current when switching unloaded high-voltage transmission lines. The actual construction of the cable is of importance when we want to study the transient recovery voltage in detail. For three-phase cables with the three conductors in a lead sheet with an earth screen, the same coupling effects as with transmission lines occur.

3.2.6. Lightning-Induced Transients

Overvoltages in a power system can be caused by transient currents and by transient voltages after switching actions during normal operation or after clearing fault situations. The overvoltages originate from the state of the system. There are also overvoltages that come from outside the system as a result of atmospheric discharges. Large parts of the power system are formed by overhead transmission lines interconnected by outdoor substations. Only in densely populated areas, the high-voltage transmission and distribution is done with high-voltage cables interconnected by gas-insulated substations (GIS) placed in buildings. When we realise that on an average every commercial aeroplane and every square kilometre of the earth's surface in a country like India is hit by lightning once a year, it is obvious that our power systems should be protected against lightning strokes. For the analysis of the lightning-induced overvoltages, a difference is made between the following :

- Lightning strokes in the vicinity of high-voltage transmission lines, which do not hit the conductors themselves,
- Direct lightning strokes on the line conductors injecting a current wave on the line, and
- Lightning strokes on the transmission towers or on the ground wires.

In addition, the solar system can be the cause of a blackout of the power system. Heavy eruptions on the solar surface in a so-called solar spot can cause a large blackout in grid in. Geomagnetic-induced currents were the cause of a DC component in the system current, and the power transformers were brought into saturation. The system current got strongly distorted and contained, apart from the power frequency ground wave, higher harmonics. The higher harmonics increased the apparent reactance of the power transformers, and therefore the transformer also consumed more reactive power. The temperature of the transformer went up, and after a certain period of time, the transformer protection tripped the power transformer out of service.

3.2.6.1. The Mechanism of Lightning

Lightning mostly occurs on summer days when the ambient temperature is high and the air is humid. Because of the temperature difference, the humid air is lifted to higher altitudes with a considerable lower ambient temperature. Cold air contains less water than warm air and raindrops are formed. The raindrops have a size of a few millimeters and are polarised by the electric field that is present between the lower part of the ionosphere and the earth's surface. The strength of this atmospheric field is on summer days in the order of 60 V/m and can reach values of 500 V/m on a dry winter day. The vertical movement of the raindrops and the wind shear splits the raindrops into negatively charged small drops and positively charged larger raindrops. The larger raindrops fall, under the influence of the gravity, to earth and create a shower. The positive charge of raindrops in a shower is confirmed by measurements. The majority of the thunderclouds is negatively charged with a potential to earth of several hundreds of megaVolts. The clouds move at great heights and the average field strength is far below the average breakdown strength of air. Inside the thundercloud, the space charge formed by the accumulating negative raindrops creates a locally strong electric field in the order of 10000 V/m and accelerates the quickly moving negative ions to considerable velocities. Collision between the accelerated negative ions and air molecules, creates new negative ions, which on their part are accelerated, collide with air molecules and free fresh negative ions. An avalanche takes place, and the space charge and the resulting electric field grow in a very short period of time. The strong electric field initiates discharges inside the cloud, and a negative stream of electrons emerges as a dim spark called a stepped leader or dart leader that jumps in steps of approximately 30 meters and reaches the earth in about 10 milliseconds. The stepped leader reaches close to the earth's surface, reaching an upward positive leader, and forms the main channel. Because of the stochastic behaviour of the space charge accumulation, the stepped leader also creates branches to the main channel. The main channel carries initially a discharge current of a few hundred Amperes having a speed of approximately 150 km/s. This discharge current heats up the main channel and the main discharge, the return stroke, is a positive discharge, and it travels at approximately half the speed of light equalizing the charge difference between the thundercloud and the earth. The main discharge current can be 100 000 A or more and the temperature of the plasma in the main channel can reach values as high as 30 000 K. The pressure in the main channel is typically 20 bar. The creation of the return stroke takes place between 5 and 10 μs and is accompanied by a shock wave that we experience as thunder. A lightning stroke consists of several of these discharges, usually three or four, with an interval time of 10–100 μs. The human eye records this as flickering of the lightning and our ear hears the rolling of the thunder. After each discharge, the plasma channel cools down to approximately 3000 K, leaving enough ionization to create a new conducting plasma channel for the following discharge. In the majority of the cases, a thundercloud is negatively charged, but positively charged clouds can also be formed.

Cloud-to-ground strokes can hit substations, transmission line towers, and transmission lines directly, but a considerable number of atmospheric discharges are between clouds. When the charged clouds are floating above, for instance, a high-voltage transmission line, they induce charge accumulation on the line conductors (see Fig. 13). When the lightning stroke equalises the charge difference between the clouds, the rather slowly accumulated charge on the conductors has to disappear at once. This results in transient currents and overvoltages.

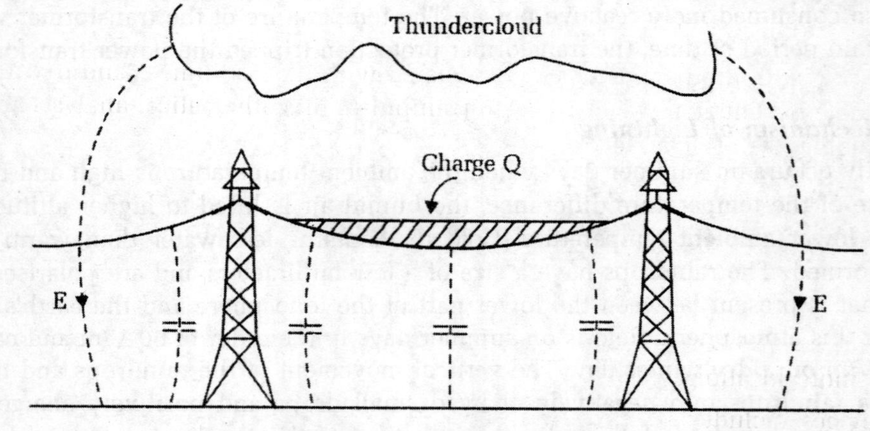

Fig. 13. *Effect of a charged cloud on a high-voltage transmission line.*

3.2.6.2. Waveshape of the Lightning Current

Lightning currents differ in amplitude and shape. The majority of the cloud-to-ground lightning strokes vary from kiloAmperes to several tenths of kiloamperes. Strokes above 100 000 amperes are rare, and the highest reported peak value of the return stroke current is 200 000 A. The shape of the current wave and the related voltage wave is rather capricious and different for every stroke. To facilitate testing in the laboratory and computations either by hand or by computer, the shape of the current wave of the return stroke is standardised. The IEC has standardised the so-called 1.2/50-μs waveform (Fig. 14).

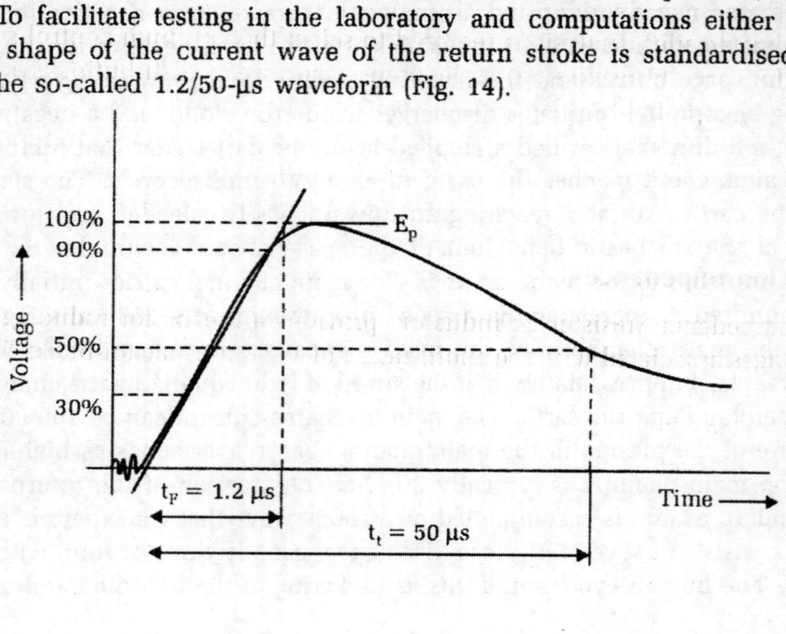

The rise time t_f = 1.2 μs is defined as being 1.67 times the time interval between 30 percent and 90 percent of the peak value of the current wave. The tail value t_t = 50 μs is defined as the time it takes until the wave drops till 50 percent of the peak value. When the current wave travels in the power system, there is, of course, a related voltage wave also present. The ratio between the voltage wave and the current wave at a certain place in the system is the characteristic impedance at that particular part of the network. System components can be exposed to very high lightning-induced overvoltages. The name plate of high-voltage equipment shows the Basic Insulation Level or BIL, which is a standardised figure for each voltage rating related to the voltage level at which the equipment should operate.

The lightning-induced voltage wave can be described mathematically as the difference of two exponential functions :

$$e(t) = E_p(e^{-\alpha t} - e^{-\beta t})$$

In this expression, the parameter β is associated with the rise time t_f and α with the tail time t_t. The double exponential wave is easy to manipulate in mathematical analysis and results in an acceptable degree of accuracy.

3.3. OVERVOLTAGE MITIGATION TECHNIQUES

The devices currently available for transient overvoltage control either attempt to minimize the transient overvoltage (or overcurrent) at the point of application, or limit (clip) the overvoltage at local and remote locations.

These devices include :

- Pre-insertion resistors (transmission and distribution)
- Pre-insertion inductors (transmission)
- Synchronous closing (transmission and distribution)
- Fixed inductors (transmission and distribution)
- MOV arresters (transmission and distribution).

Previous studies have suggested that the effectiveness of these control methods is system dependent, and that detailed analysis is required to select the optimum control scheme. While often justifiable for large transmission applications, analysis of distribution system capacitor applications is rarely completed, and in general, banks are installed without transient overvoltage control. Each of these methods has various advantages and disadvantages in terms of transient overvoltage reduction, cost, installation requirements, operating/maintenance requirements, and reliability.

3.3.1. Pre-insertion Impedance

A pre-insertion impedance (resistor or inductor) provides a means for reducing the transient currents and voltages associated with the energization of a shunt capacitor bank. The impedance is "shorted-out" (bypassed) shortly after the initial transient dissipates, thereby producing a second transient event. The insertion transient typically lasts for less than one cycle of the system frequency. The performance of pre-insertion impedance is evaluated using both the insertion and bypass transient magnitudes, as well as the capability to dissipate the energy associated with the event, and repeat the event on a regular basis.

Pre-insertion resistors are one of the most effective means for controlling capacitor energizing transients; however, reliability issues have often caused utilities to select other means. The optimum resistor value for controlling capacitor energizing transients depends primarily on the capacitor size and the source strength. It should be approximately equal to the surge impedance (Z_S) formed by the bank and source :

$$R_{optimum} \approx \sqrt{(L_S/C)}$$

Pre-insertion inductors, which are primarily used for overcurrent control for back-to-back applications also provide some level of transient overvoltage reduction. Often applied to circuit switchers, inductors are less sensitive to thermal considerations (resistor failure mode) and are more economical than resistors.

3.3.2. Synchronous Closing Control

Synchronous closing is independent contact closing of each phase near a voltage zero. To accomplish closing at or near a voltage zero (avoiding high prestrike voltages), it is necessary to apply a switching device that maintains a dielectric strength sufficient to withstand system voltages until its contacts touch. Although this level of precision is difficult to achieve, closing consistency of ±0.5 milliseconds should be possible. Previous studies have indicated that a closing consistency of ±1.0 millisecond provides overvoltage control comparable to properly sized pre-insertion resistors. The success of a synchronous closing scheme is often determined by the ability to repeat the process under various (system and climate) conditions.

Ungrounded banks are controlled by closing the first two phases at a phase-to-phase voltage zero and then delaying the third phase 90 degrees (phase-to-ground voltage zero).

3.3.3. Fixed Inductors

Fixed inductors have been used successfully to limit inrush currents during back-to-back switching. Typically the value of these inductors is on the order of several hundred micro-henries. In addition, inductors provided for outrush (into a nearby fault) current control may be applied, and are typically 0.5 – 2.0 mH. Previous simulations indicate that these fixed reactors do not provide any appreciable transient overvoltage reduction.

3.3.4. MOV Arresters

Metal oxide varistors (MOV's) can limit the transient voltages to the arrester's protective level (maximum switching surge protective level, typically 1.8 – 2.5 per unit) at the point of application. The primary concern associated with MOV application is the energy duty during a restrike event. Although a rare occurrence, a switch restrike generally results in the highest arrester duty for arresters located at the switched capacitor. In addition, remote arresters (including low voltage customer applications) may be subjected to severe energy duties if voltage magnification occurs. This condition could be especially troublesome for distribution systems if SiC arresters remain in service.

3.3.5. Application Experience

Each of the technologies previously described has been utilized in the field with varying degrees of success. The criteria by which these devices are evaluated, however, is changing significantly. For example, design requirements often state that protection of utility equipment (*i.e.*, transformers—phase-to-phase transients) is the primary factor. However, the recent concern for customer systems has prompted a number of utilities to seek a "transient free" solution. This fact is best illustrated by reviewing the concern for nuisance tripping of adjustable-speed drives. As previously stated, the dc bus overvoltage trip setting for small voltage-source inverter drives may be as low as 117% of nominal, using this value as the maximum (low voltage) transient during utility capacitor switching means that the mitigation technology will have to be very "robust". A recent study indicated that even with an error window of ±1.0 millisecond (using synchronous closing control), the dc bus levels would be within 10% of the drive's trip setting (drive without input choke).

In addition to the overvoltage design limits, there are a number of other factors that have delayed the widespread application of mitigation technologies. One obvious obstacle is cost, however, an equally important reason has been reliability. Several examples include :

- A number of utilities indicate that the cost of the technology has not reached a point where widespread application is feasible, especially at the distribution level where the number of banks "near" sensitive customers is significant. Another factor in the cost is system studies, and custom designs (utilities would prefer to apply one technology and have it work anywhere on the system).
- Pre-insertion resistor failures (thermal) prompted one manufacturer to move to a pre-insertion inductor design. Several studies have indicated that under certain system conditions (*i.e.*, weak source), these inductors provide minimal local overvoltage control and may even cause the remote overvoltages to increase.
- A number of utilities have reported drift (usually climate related) of the synchronous closing control. Even a minor error could be enough to produce low voltage transients sufficient to cause drive tripping.

  The error is correctable, however, it requires a technician to "zero-in" the controls.
- A number of utilities have reported equipment failure during a capacitor switch restrike event. Although the mitigation scheme does not play a role in the restrike, a lack of arresters (MOV's) due to the existence of the control, would certainly affect the event (perhaps leading to multiple restrikes and subsequent equipment failure). These, and most certainly other factors have prompted a number of utilities to evaluate elaborate schemes, and have motivated manufacturers to design new options.

Customer solutions for problems related to utility capacitor switching transients include chokes, low voltage arresters, and harmonic filters. Although there are other power conditioning options, such as TVSS, the options listed below provide the best protection from the voltage magnification and nuisance tripping phenomena.

3.3.6. Chokes

A choke (or reactor, or isolation transformer) may be used to provide isolation for a power electronic device, such as an adjustable-speed drive (as shown in Fig. 15). In general, these devices are low cost (relative to cost of drive) and easy to install. Several manufacturers now include them as a standard option, primarily to reduce harmonic current levels and solve the nuisance tripping problem. Care should be taken when selecting an appropriate size for the input choke. If the size of the choke is too large, the drive may not function properly due to the voltage drop. Typically, standard sizes of 1.5%, 3%, and 5% are used to solve the nuisance tripping problem—with 5% being the upper limit.

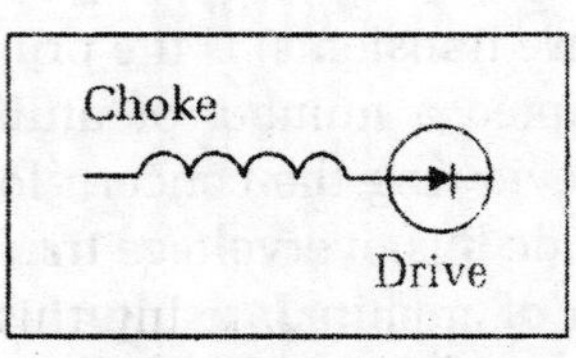

Fig. 15. *Typical choke/drive configuration.*

3.3.7. Arresters

Low voltage (440 volt) arresters may be used to limit overvoltages to acceptable levels. For low voltage applications, the energy rating of the arrester is the critical consideration. Small, inexpensive, MOV's that are found in TVSS and power electronics typically have energy ratings of only several hundred joules (or less).

Simulations indicate that for the voltage magnification case, arrester duties may exceed several thousand joules. Therefore it is necessary to install "high-energy" MOV's for this application. These MOV's are modeled using an arrester maximum switching surge protective level (MSSPL) of 700 volts (based on a 500 Amp surge). Fig. 16 illustrates the arrester protective level and energy capability of one particular brand of low voltage arrester (480 V).

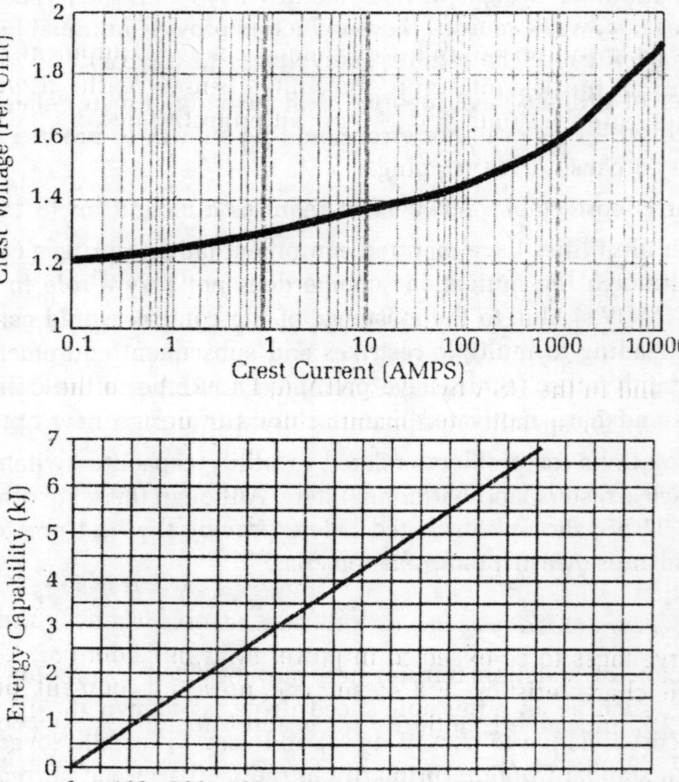

Fig. 16. *Low voltage arrester protective level and energy capability.*

3.3.8. Filters

Harmonic filters are often applied to reduce harmonic voltage distortion levels. Since power factor correction capacitors are the key element in the voltage magnification case, it is worthwhile to evaluate the impact of the filter inductor on the transient voltage. In most cases, converting a power factor correction bank into a harmonic filter will reduce the transient overvoltage due to :

1. Adding the filter inductor changes the resonance frequency. Typically, this will cause the two frequencies to move further apart.
2. In addition, if magnification still occurs, it will be across the capacitor (and not the bus). A significant amount of the transient voltage will be dropped across the inductor, thereby improving the bus voltage.

3.4. TRANSIENT OVERVOLTAGES IN COMMUNICATION CIRCUITS

Communication circuits share with power circuits their exposure to lightning effects, but differ from power circuits in that they do not have the power switching effects just discussed. On the other hand, they are exposed to another source of overvoltages—the injection of power-frequency current into the circuits. This injection can be the result of accidental contact between fallen power wires on overhead communication wires or cables, or even a malicious injection of power-frequency voltage into terminals of communication equipment. This injection can last until a power circuit breaker interrupts the fault it senses on the power system, or it can be enduring, in which case the communication circuit protective devices must provide an acceptable failure mode to maintain the safety and integrity of the communication plant - with interruption of the service until the fault condition is removed.

3.5. STANDARDS ON TRANSIENT OVERVOLTAGES

Several standards or guides have been issued or proposed in Europe by VDE, IEC, CECC, Pro-Electron, CCITT, and in the USA by IEEE, NEMA, UL, REA and the military, specifying a surge withstand capability for specific equipment or devices and specific conditions of transients in power or communication systems. Some of these specifications represent early attempts to recognize and deal with the problem in spite of insufficient data. As a growing number of organizations address the problem and as exchanges of information take place improvements are being made in the approach. The IEEE has recently published a new standard describing the environment in low-voltage ac power circuits. The Low-Voltage Insulation Coordination Subcommittee SC/28A of IEC has also completed a report IEC 664, listing the maximum values of transient overvoltages to be expected in power systems under controlled conditions and for specified system characteristics. For some time now, a document prepared by a Relaying Committee of IEEE under the title Surge Withstand Capability has been available. These three documents are reviewed in the pages that follow. Greater emphasis will be placed on the first cited above because it describes the transient environment : the other two assume an environment for the purpose of specifying tests.

3.5.1. The IEEE Surge Withstand Capability Test

It was designed to address new problems in electronic equipments exposed to power system transients was prepared by an IEEE committee dealing with the exposure of power system relaying equipment to the harsh environment of high-voltage substations. This describes a transient generated by the arcing that takes place when air-break, disconnect switches are opened or closed in the power system presents innovations in transient protection :

- The specified waveshape of the voltage is oscillatory wave-shape, not the historical unidirectional waveshape.
- A source impedance, a characteristic undefined in many other documents is defined.
- The concept that all lines to the device under test must be subjected to the test is spelled out.

Because this useful test was released at a time when little other guidance was available, users attempted to apply the recommendations of this test to situations where the environment of a high-voltage substation did not exist. Thus, an important consideration in the application of tests dealing with transients is a clear definition of the scope, and limitations of the application.

3.5.2. The IEEE Guide on Surge Voltages (IEEE Std 587–1980)

Voltages and Rate of Occurrence

Data collected from a number of sources led to plotting a set of lines representing a rate of occurrence as a function of voltage for three types of exposures. These exposure levels are defined in general terms as follows :

- Low Exposure—Systems in geographical areas known for low lightning activity, with little load switching activity.
- Medium Exposure—Systems in geographical areas known for high lightning activity, with frequent and severe switching transients.
- High Exposure—Rare but real systems supplied by long overhead lines and subject to reflections at line ends, where the characteristics of the installation produce high sparkover levels of the clearances.

It is essential to recognize that a surge voltage observed in a power system can be either the driving voltage or the voltage limited by the sparkover of some clearance in the system. Hence, the term unprotected circuit must be understood to be a circuit in which no low-voltage protective device has been installed but one in which clearance sparkover will eventually limit the maximum voltage. The distribution of surge levels, therefore, is influenced by the surge-producing mechanisms as well as by the sparkover level of clearances in the system. This distinction between actual driving voltage and voltage limited by sparkover is particularly important at the interface between outdoor equipment and indoor equipment. Outdoor equipment has generally higher clearances, hence higher sparkover levels : 10 kV may be typical, but 20 kV is possible. In contrast, most indoor wiring devices used in 120 to 240 V systems have sparkover levels of about 6 kV; this 6 kV level, therefore, can be selected as a typical cutoff for the occurrence of surges in indoor power systems.

The voltage and current amplitudes presented in the Guide attempt to provide for the vast majority of lightning strikes but none should be considered "worst case," as this concept cannot be determined realistically. It is necessary to think in terms of the statistical distribution of strikes, and to accept a reasonable upper limit for most cases. Where the consequences of a failure are not catastrophic but merely represent an annoying economic loss, it is appropriate to make a tradeoff of the cost of protection against the likelihood of a failure caused by a high but rare surge.

3.5.3. Energy and Source Impedance

The energy involved in the interaction of a power system with a surge source and a surge protective device will divide between the source and the protective device in accordance with the characteristics of the two impedances. In a gap-type protective device, the low impedance of the arc after sparkover forces most of the energy to be dissipated elsewhere—for instance, in a resistor added in series with the gap for limiting the power-follow current. In an energy-absorber protective device, by its very nature, a substantial share of the surge energy is dissipated in the suppressor, but its clamping action does not involve the power-follow energy resulting from the short-circuit action of a gap. It is therefore essential to the effective use of surge protective devices that a realistic assumption be made about the source impedance of the surge whose effects are to be duplicated.

Unfortunately, not enough data have been collected on what this assumption should be for the source impedance of the transient Standards and recommendations, such as MIL STD-1399 or the IEC 664 report, either ignore the issue or indicate values applicable to limited cases, such as the SWC test for high-voltage substation equipment. The IEEE 587 standard attempts to relate impedance to categories of locations but unavoidably remains vague on their definitions.

The 6 kV open-circuit voltage derives from two facts : the limiting action of wiring device sparkover, and the unattenuated propagation of voltages in unloaded systems. The 3 kA discharge current in Location Category B derives from experimental results : field experience in surge protective device performance and simulated lightning tests. The two levels of discharge currents for the 0.5 μs–100 kHz wave derive from the increasing impedance expected in moving from Category B to Category A.

Category C is likely to be exposed to substantially higher voltages than Category B because the limiting effect of sparkover might not be available. The high exposure rates could apply, with voltage in excess of 10 kV and discharge currents of 10 kA or more. Installing unprotected load equipment in Category C is not recommended; the installation of secondary arresters, however, can provide the necessary protection.

Having defined the environment for low-voltage ac power circuits, the Working Group is now preparing an Application Guide, where a step-by-step approach will outline the method for assessing the need for transient protection and selecting the appropriate device or system. Parallel work in other IEEE working groups preparing test specification standards for surge protective devices will be helpful in this selection process. Other groups in the U.S. as well as the international bodies of IEC and CCITT are now working toward further refinements and the reconciliation of different approaches.

3.6. TRANSIENT MEASUREMENTS

Transient measurements in the laboratory are easier to obtain than are transient measurements for monitoring purposes, but precautions are nevertheless necessary—technical precautions to ensure the accuracy of the experiment and safety precautions to ensure the survival of the experimenter.

The fundamental tool for measurements is still the cathode—ray oscilloscope, although digital instruments are being improved in performance and convenience. Great advances over the cumbersome taking of photographs with a non-persistent oscilloscope display have been made in fast-storage oscilloscopes that speed the examination of test data. Major oscilloscope manufacturers offer a wide choice of instruments, and their catalogues and application information are valuable for selecting the correct instrument. Where fast transients are involved, it is useful to remember that minor secondary oscillations, which may not be of any consequence to the circuit performance, tend to produce a very fast sweep of the electron spot, often making the trace invisible except at the peaks of the secondary oscillation and producing a dotted line trace rather than the clean record expected in principle. It is the clean oscillograph record that is valuable in understanding phenomena and that makes the oscilloscope the best friend (if trustworthy) of the engineer.

This trustworthiness of the oscilloscope can and should be tested by the engineer before he draws hasty conclusions from transient measurements. Here again the literature will provide guidance; in addition there are some simple checks : freedom from background noise and radiation into the oscilloscope preamplifiers, alertness to the pitfalls of the built-in delay line, proper compensation of attenuator probes, and avoidance of ground loops.

When repetitive transient tests are involved, digital instruments are valuable and require less operator training. For large evaluation programs, incoming or outgoing quality control, automated equipment is now available, either as separate monitors or in combination with the surge generators.

3.7. TRANSIENT (SURGE) GENERATORS

The basic surge generator is generally a charged capacitor, discharged through a wave-shaping circuit into the test piece. A variant can be a pulse-forming network providing a quasi-constant pulse form, generally of the flat-top variety. In some cases involving a current source, special circuits have been developed to provide constant-current, switchable power supplies.

Since most surge tests are based on some agreed upon standard, there are generally only a limited number of test waveshapes to be produced for a given application. Ideally, the surge generator should be so powerful that undesirable loading by the test piece could not occur. In some cases, however, loading of the circuit by the interaction of the generator source impedance and the test piece impedance is precisely the point of the experiment. Practical limitations also exist on the cost of a large-capacity machine, the difficulty in maintaining fast fronts for high-power impulses, and the safety hazards of large energy storage.

Some test standards include in their specifications the test circuit to be used; in that case the engineer has little choice except to build the circuit or to find a commercial equipment supplier of the gear. Other standards specify only the electrical characteristics of the impulse, leaving the exact test circuit to be determined by the laboratory engineer.

Traditionally, manufacturers of utility equipment have maintained competent staffs of laboratory engineers and technicians to perform the required tests. Such finely honed competence is not always found in the laboratories of electronics manufacturers, in part because the need has not been as obvious as it is in the high-voltage equipment. While it may be highly educational for the electronic engineer to build his own surge generator, the process can be frustrating and dangerous if safety precautions are not rigidly observed because many capacitors are not simply capacitors but have internal inductance, and inductors have stray capacitance. Instead of obtaining test results on his equipment, the busy engineer observes overshooting, ringing, and exploding carbon resistors. Thus, for purely economic reasons, the typical electronics engineer would be better advised to purchase a commercial surge generator, even if it is custom made. These are the reasons that an equipment demonstration has been included in the program.

3.8. TRANSIENT SUPPRESSORS

Two methods and types of devices are available to suppress transients : blocking the transient through some low-pass filter, or diverting it to ground through some nonlinear device. This nonlinearity may be either a frequency nonlinearity (high-pass filter) or a voltage nonlinearity (clamping action or crowbar action). We will be mostly discussing the second type in this paper, since voltage clamping devices or crowbar devices are the most frequently used.

Voltage-clamping devices have variable impedance, depending on the current flowing through the device or the voltage across its terminal. These components show a nonlinear characteristic, *i.e.*, Ohm's law can be applied but the equation has a variable R. Impedance variation is monotonic and does not contain discontinuities, in contrast to the crowbar device which shows a turn-on action. As far as volt-ampere characteristics of these components is concerned, they are time-dependent to a certain degree. However, unlike sparkover of a gap or triggering of a thyristor, time delay is not involved here.

When a voltage-clamping device is installed, the circuit remains unaffected by the device before and after the transient for any steady-state voltage below clamping level. Increased current drawn through the device as the voltage attempts to rise results in voltage clamping action. Nonlinear impedance is the result if this current rise is faster than the voltage increase. Increased voltage drop (IR) in the source impedance due to higher current results in the apparent clamping of the voltage. It should be emphasized that the device depends on the source impedance to produce the clamping. A voltage divider action is at work where one sees the ratio of the divider not constant, but changing. The ratio is low, however, if the source impedance is very low. The suppressor cannot work at all with a limit zero source impedance. In contrast, a crowbar type device effectively short-circuits the transient to ground; but once established, this short circuit will continue until the current (the surge current as well as any power-follow current supplied by the power system) is brought to a low level.

The crowbar device will often reduce the line voltage below its steady-state value, but a voltage clamping device will not. Substantial currents can be carried by the suppressor without dissipating a considerable amount of energy within the suppressor, since the voltage (arc or forward-drop) during the discharge is held very low. This characteristic constitutes the major advantage of these suppressors. However, limitations in volt-time response and power-follow are the price paid for this advantage. As voltage increases across a spark-gap, significant conduction cannot take place until transition to the arc mode has taken place by avalanche breakdown of the gas between the electrodes. The load is left unprotected during the initial rise due to this delay time (typically in microseconds). Considerable variation exists in the sparkover voltage achieved in successive operations, since the process is statistical in nature. This sparkover voltage can, in addition, be substantially higher after a long period of rest than after successive discharges, for some devices. From the physical nature of the process, it is difficult to produce consistent sparkover voltage for low voltage ratings. This difficulty is increased by the effect of manufacturing tolerances on very small gap distances. This difficulty can he alleviated by filling the tube with a gas having lower breakdown voltage than air. However, if the enclosure seal is lost and the gas replaced by air this substitution creates a reliability problem because the sparkover of the gap is then substantially higher.

Another limitation occurs when a power current from the steady-state voltage source follows the surge discharge (follow-current, or power-follow). This power-follow current may or may not be cleared at a natural current zero, in ac circuits fit dc circuits, clearing is even more uncertain. Additional means must, therefore, be provided to open (lie power circuit, if the crowbar device is not designed to provide self-clearing action within specified limits of surge energy, system voltage, and power-follow current. Consequently, most electronic circuits are better protected with voltage clamping suppressors than with crowbars, but sometimes the energy deposited in a voltage clamping device by a high current surge can be excessive; a combination of the two devices can provide an effective protection at optimum cost. However, this combined protection must be properly coordinated to obtain the fall advantage of the scheme.

3.9. INTERRUPTION OF FAULT CIRCUITS

In the 1950s, it was observed that minimum-oil circuit breakers, in particular, exploded after clearing short-circuit currents with a current level below the nominal rated short-circuit breaking capability. In addition, air-blast breakers occasionally failed to interrupt short-circuit currents considerably smaller than the current values they were designed and tested for. Careful analysis has revealed that in many cases the fault occurred on a high-voltage transmission line, a few kilometres from the circuit breaker terminals. The short-line fault was discovered, and several years later the short-line fault test appeared in the standards. Presently, the short-line fault test is considered to be one of the most severe short-circuit duties for high-voltage circuit breakers. The short-line fault test creates, after current interruption, a very steep voltage oscillation at the line side of the breaker. This puts a high stress on the still-hot arc channel, and this can cause a thermal breakdown of the arc channel. The arcing time is prolonged, and when the same happens at the next current zero, the breaker fails to interrupt the short-circuit current. SF6 circuit breakers are especially very sensitive for short-line faults occurring only a few hundred metres to a few kilometres from the breaker terminals. Air-blast breakers are, in particular, stressed by faults occurring further away, several kilometers from the breaker terminals. The reason for this will be explained later in this paragraph; first the TRVs at the supply side and later the TRVs at the line side of the breaker will be analysed. Fig. 17 shows the single-phase representation of a short-line fault.

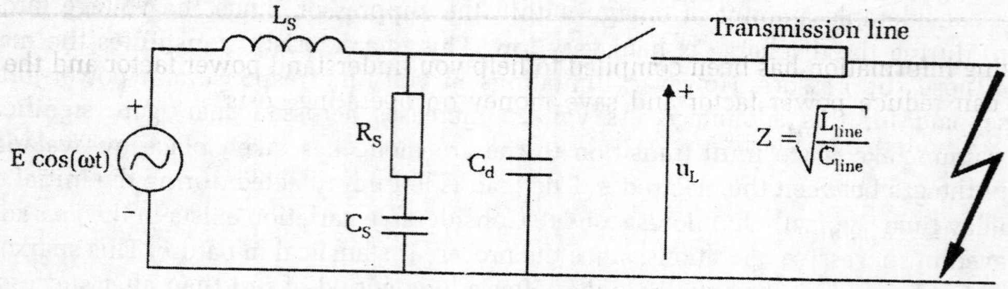

Fig. 17. *The short-line fault.*

After interruption of the short-line fault current, the TRV builds up across the breaker contacts. This TRV consists of a voltage generated by the supply network and of a voltage created by the line-side oscillation. The TRV from the supply network has a more or less (1-cosine) shape, which we can expect because it is basically an oscillating LCR-series network. The line-side oscillation has a triangular-shaped or saw-tooth-shaped waveform.

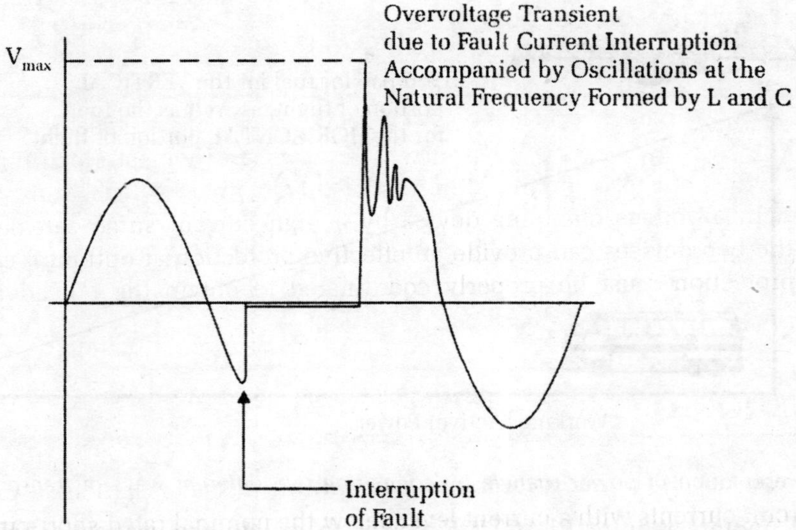

Fig. 18. *Overvoltage transient due to fault current interruption.*

The circuit breaker interrupts the short-circuit current at current zero. The rms value of the current is less than the current in the case of a bolted terminal fault directly behind the terminals of the breaker. The impedance of the short-circuited length of line reduces the current. Because of the inductive nature of the circuit, the supply voltage $E \cos(\omega t)$ is close to its peak value, and the voltage at the line side has the value $u_L = L_{Line}\ di/dt$.

The voltage profile over the line decreases linearly from u_L to zero, when we consider a single line-to-ground fault, and the line is therefore left charged at the instant of current interruption. The charge is maximal at the side of the breaker and zero at the fault place. The charge on the line will balance itself and this balancing of charges causes travelling waves on this part of the high-voltage transmission line. The reflections of the travelling waves against the short-circuit point and the open end at the breaker side cause a triangular-shaped waveform.

3.10. POWER FACTOR CORRECTION USING CAPACITORS

The following information has been compiled to help you understand power factor and the way capacitors can reduce power factor and save money on operating costs.

This is divided into the following sections :

- Power factor background
- Installing capacitors

3.10.1. Background

Power factor is an electrical ratio. The power factor compares the capacity for work produced by the system to the amount of power going into a system. Consideration of power factor can be an important element in reducing operating costs.

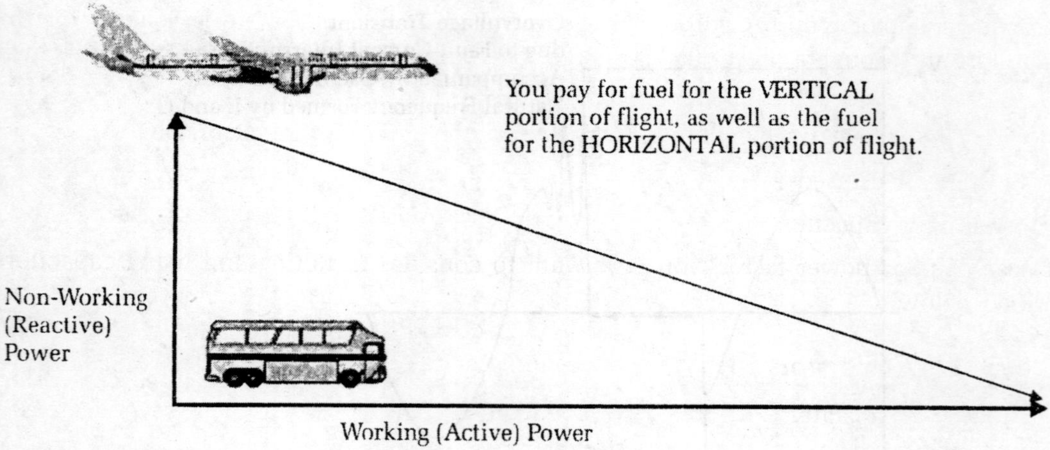

Fig. 19. *Representation of power triangle by comparing two different ways of transportation.*

- Similarly, motors require REACTIVE power to set up the magnetic field while the ACTIVE power produces the useful work (shaft horsepower). Total Power is the vector sum of the two and represents what you pay for :

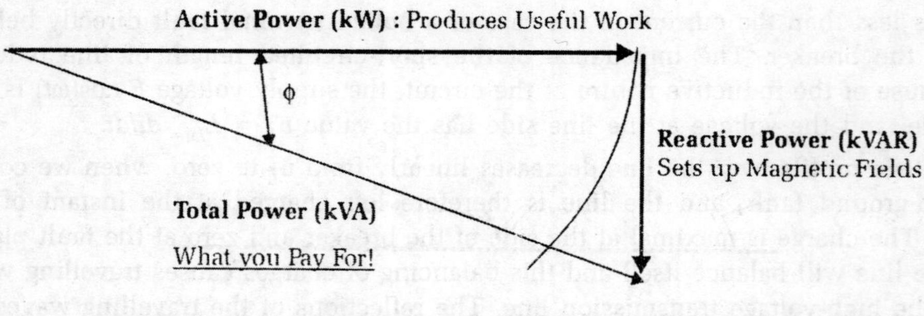

Fig. 20. *Relationship between different components of power.*

- Power Factor is the ratio of Active Power to Total Power :

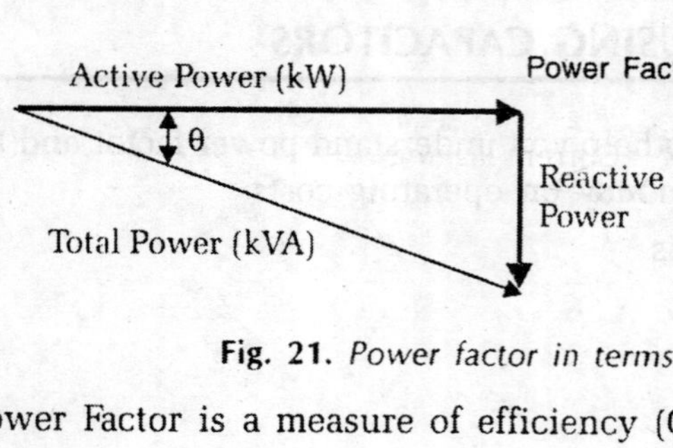

$$\text{Power Factor} = \frac{\text{Active (Real) Power}}{\text{Total Power}} = \frac{\text{kW}}{\text{kVA}} = \text{Cosine } (\theta)$$

Fig. 21. *Power factor in terms of different parameters.*

- Power Factor is a measure of efficiency (Output/Input).

A high power factor indicates that the electricity powering a piece of equipment is used efficiently. A high quantity and concentration of wire windings within a piece of equipment reduces the power factor. Equipment with fewer wire windings, such as light bulbs or electric resistance heaters, has a higher power factor ratio than does a motor, which contains a great deal of wire windings.

Poor power factor requires additional current to serve the load. Utilities charge for this additional current through a power factor charge. There are three basic ways to improve power factor. They are :

- using the correct sized motor for a particular piece of equipment or job
- shutting off unused motors, and
- installing a capacitor.

In case of poor power factor, you may want to consider installing and using capacitors, as described below.

3.10.2. Installing Capacitors

Before installing a capacitor, a facilities manager should conduct research in several areas.

- Conduct a harmonics study, particularly in environments with a large amount of variable speed equipment. This will ensure a proper capacitor application.
- Consult the manufacturer of any large equipment to which you will attach a capacitor to determine the manufacturer's specifications. This will assure that the installation will reduce operating costs while maintaining product quality and production levels.
- Work closely with Power engineers to ensure system capability.

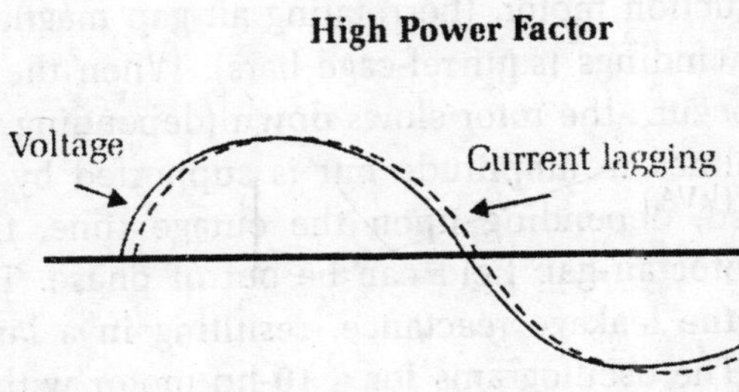

Fig. 22. *Waveform shown is of high power factor.*

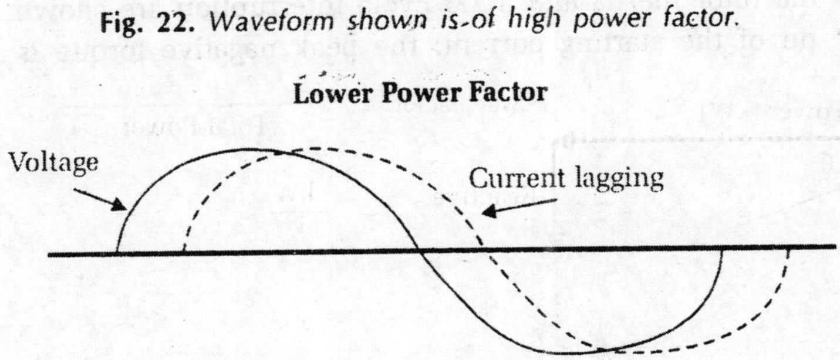

Fig. 23. *Waveform shown is of lower power factor.*

Capacitors increase the power factor, letting the maximum current occur closer to the time when the voltage is at a maximum. A capacitor may be thought of as a battery that stores and releases current to improve the power factor.

- Capacitors supply, for free, the reactive energy required by inductive loads.
  - You only have to pay for the capacitor
  - Since the utility doesn't supply it (kVAR), you don't pay for it

Fig. 24. *Installing a capacitor.*

- Voltage Improvement :
  - When capacitors are added, voltage will increase
  - Typically only a few percent
  - Not a significant economic or system benefit.

3.11. MOTOR START TRANSIENT

In the normal operation of the induction motor, the rotating air-gap magnetic field links both the stator windings and the rotor windings (squirrel-cage bars). When the stator voltage sags or is interrupted, two phenomena occur : the rotor slows down (depending on the load inertia) and the air-gap magnetic field declines in amplitude but is supported by the rotor currents. When the stator voltage is restored, depending upon the outage time, the stator-produced magnetic field and the declining rotor air-gap field can be out of phase. The result is nearly double the line voltage acting on the leakage reactance, resulting in a large transient stator current and negative shaft torque. The oscillograms for a 10-hp motor with an inertia load of 4.28 per unit (pu) of the rotor inertia and a 7.3 cycle interruption are shown in Fig. 25. The peak current is 1.67 pu of the starting current; the peak negative torque is 1.71 pu of the starting torque.

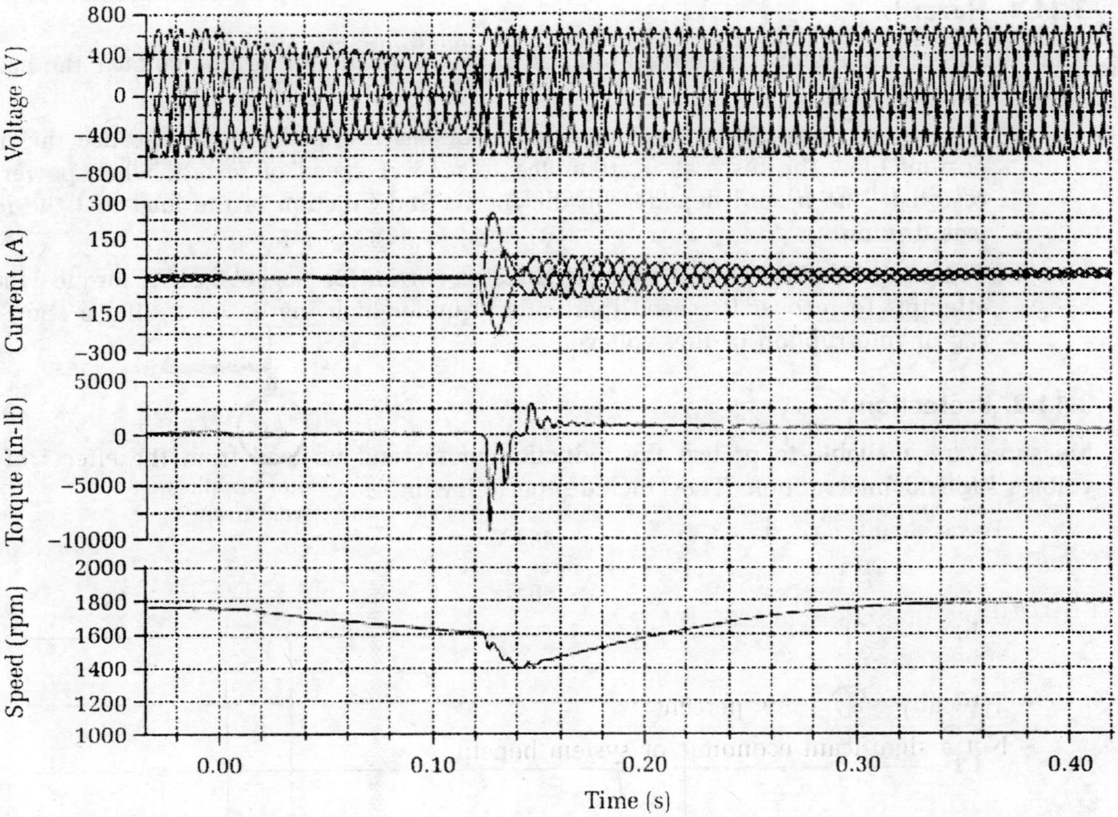

Fig. 25. *Motor voltage, current, shaft torque, shaft speed versus time for 7.3 cycle voltage interruption for a 10-hp motor.*

The impact on the induction motor of line-voltage sag is similar for an interruption of the same time duration where the motor voltage is restored. Figure 26 shows the stator current of a 7.5-hp induction motor subjected to a voltage sag to 30 percent for 5.5 cycles. The motor was loaded only with the inertia of an eddy-current brake. The current at the restoration at the end of the voltage sag is about 4.5 times rated current, of the order of the line starting current.

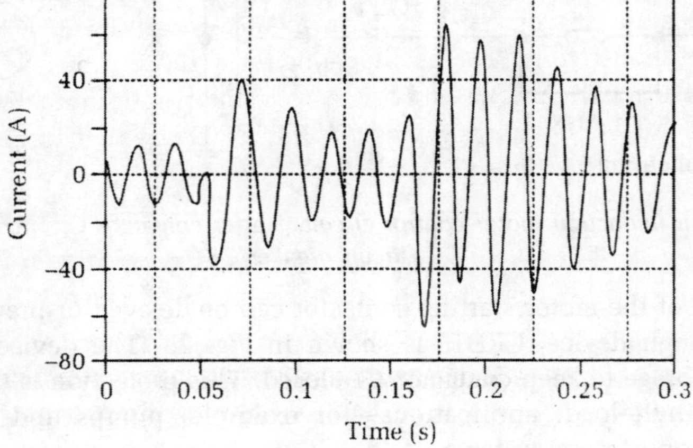

Fig. 26. *7.5-hp induction motor. Current during voltage sag to 30 percent of 5.5 cycles and after voltage is restored.*

3.11.1. Hazards

After the induction motor is subjected to a sag or total interruption of line voltage, the hazards include the following :

1. The motor stops either because a contactor has dropped out, or because the motor current blew the fuses as it attempted to recover speed or restart, when power was restored. The hazard depends on the inertia and function of the load that the motor was driving.
2. The motor damages itself from high stator current or negative shaft torque when it attempts to recover its speed (particularly under high inertia load) after a short-time sag or interruption of line voltage.

3.11.2. Protection

Measures are available to protect the induction motor and its load from the effects of line voltage sag and interruption. They include the following :

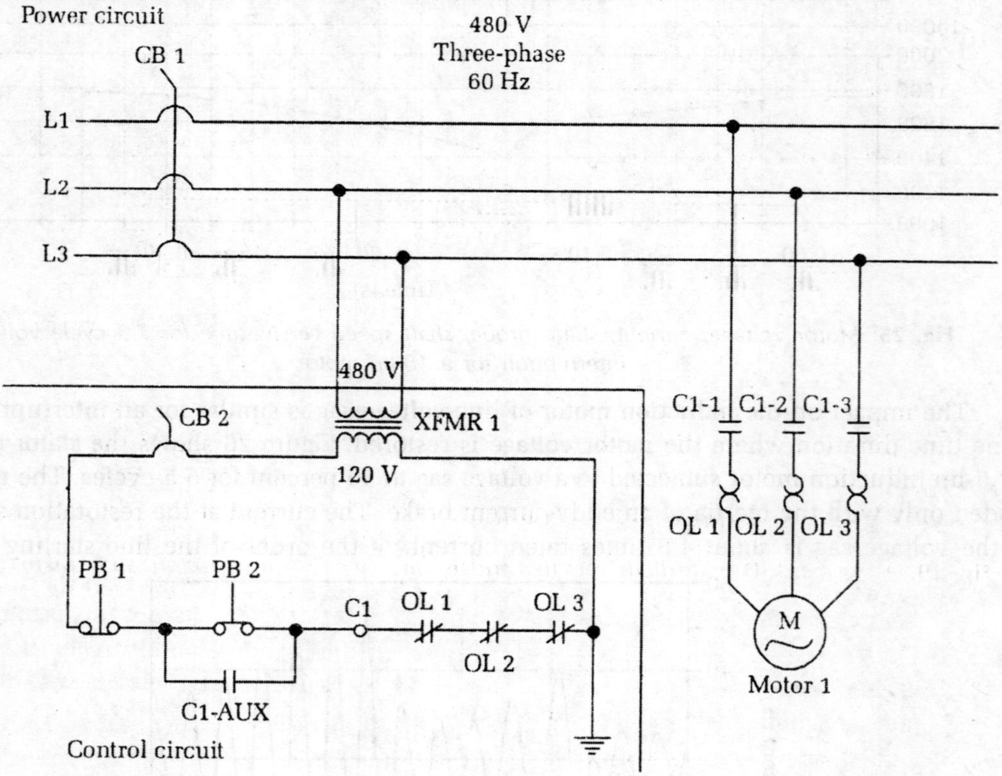

Fig. 27. *Basic induction motor control circuit. Starter contactor C, Thermal overloads OL. Circuit breaker CB.*

1. Drop out of the motor starting contactor can be delayed or prevented with a contactor ride through device, CRD1, as shown in Fig. 28. The device utilizes some type of energy storage to keep contactor *C*1 closed. This protection is most suitable for single-motor, single-load, applications—for example, pumps and fans, as compared to multimotor process systems.

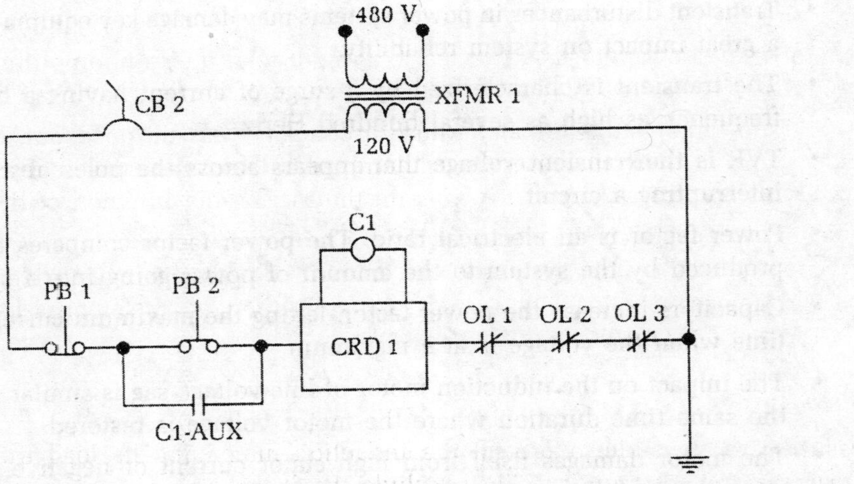

Fig. 28. *Control circuit of above figure with voltage sag contactor ride-through device CRD1.*

2. Operate the induction motor in an adjustable speed drive where the dc-link capacitor voltage is maintained during a line-voltage sag or interruption by an auxiliary source from the kinetic energy of the load.

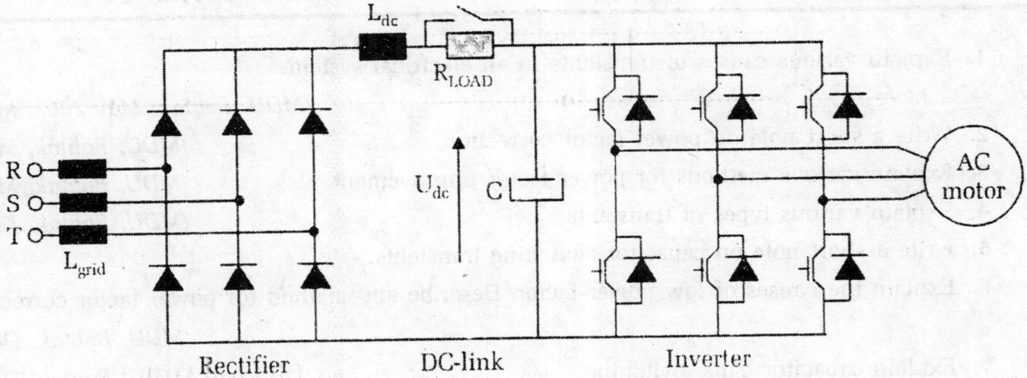

Fig. 29. *A typical ASD with diode rectifier bridge, line, and DC-link reactors and braking resistor.*

3. Control the induction motor with a microprocessor-based motor management relay. The relay will regulate restarts after voltage interruptions based on the thermal state of the motor, as compared with a count of starts per hour. The relay is suitable for large motors where the numbers of starts is limited to prevent overheating.

SUMMARY

- A transient occurs in the power system when the network changes from one steady state into another.
- An impulsive transient is a brief, unidirectional variation in voltage, current, or both on a power line.
- An oscillatory transient is a brief, bidirectional variation in voltage, current, or both on a power line.

- Transient disturbances in power systems may damage key equipment, potentially having a great impact on system reliability.
- The transient is characterized by a surge of current having a high magnitude and a frequency as high as several hundred Hertz.
- TVR is the transient voltage that appears across the poles of a circuit breaker upon interrupting a circuit.
- Power factor is an electrical ratio. The power factor compares the capacity for work produced by the system to the amount of power going into a system.
- Capacitors increase the power factor, letting the maximum current occur closer to the time when the voltage is at a maximum.
- The impact on the induction motor of line-voltage sag is similar for an interruption of the same time duration where the motor voltage is restored.
- The motor damages itself from high stator current or negative shaft torque when it attempts to recover its speed (particularly under high inertia load) after a short-time sag or interruption of line voltage.

Review Questions and Exercises

1. Explain various causes of transients in an electrical system. *(MDU, Rohtak, May 2008, May 2009)*
2. Write a short note on power factor correction. *(MDU, Rohtak, May 2009)*
3. Explain various methods for power factor improvement. *(MDU, Rohtak, May 2008)*
4. Explain various types of transients. *(MDU, Rohtak, Dec. 2008)*
5. write a short note on capacitor switching transients.
6. Explain the causes of low power factor. Describe any method for power factor correction. *(MDU, Rohtak, Dec. 2008)*
7. Explain capacitor bank switching. *(MDU, Rohtak, Dec. 2008)*

❑❑❑

Chapter 4

HARMONICS

4.1. INTRODUCTION

In the past, the most common domestic and commercial loads connected to the electricity supply system were of linear characteristics, like heater and incandescent lights. These loads do not produce harmonics, since the current of linear loads is always proportional to the applied voltage.

The issues and considerations associated with three-phase power harmonics are often misunderstood. With the advent of power electronics and proliferation of non-linear loads in industrial power applications, power harmonics and their effects on power quality are a topic of concern. Currently in India, only 15 to 20% of the utility distribution loading consists of non-linear loads. It is projected over the next ten years that non-linear loads will comprise approximately 70 to 85% of the loading on our nation's utility distribution systems.

The effects of single phase power harmonics and neutral conductors are often a source of concern. Harmonics play an important role in single phase applications if the neutral conductors are undersized, however single phase harmonics are associated with phase-to-neutral loads are not discussed in this chapter.

The characteristic harmonics are based on the number of rectifiers (pulse number) used in a circuit and can be determined by the following equation :

$$h = (n \times p) \pm 1$$

where : n = an integer (1, 2, 3, 4, 5, ...)

p = number of pulses or rectifiers

For example, using a 6 pulse rectifier, the characteristic harmonics will be :

$h = (1 \times 6) \pm 1 \Rightarrow$ 5th and 7th harmonics

$h = (2 \times 6) \pm 1 \Rightarrow$ 11th and 13th harmonics

$h = (3 \times 6) \pm 1 \Rightarrow$ 17th and 19th harmonics

$h = (4 \times 6) \pm 1 \Rightarrow$ 23rd and 25th harmonics

Fig. 1. *Determining characteristic harmonics.*

Today's electric power systems are connected to many non-linear loads. These include static power converters, arc discharge devices, and electronic control equipment by semiconductor devices, saturated magnetic devices and rotating machines. The characteristics of these non-linear loads inevitably change the sinusoidal nature of the a.c. power current, resulting in the flow of the harmonics current in the a.c. power system.

If harmonics studies are neglected, it is possible that resonance at some generated harmonics between power factor improvement capacitors and source inductance may lead to larger oscillating currents and consequently high harmonics voltage. These high voltages and currents can cause additional system losses, breakdown of equipment insulation, capacitor fuse melting, errors in metering and protective relays and interference with other consumers.

A harmonic is a component of a periodic wave having a frequency that is an integral multiple of the fundamental power line frequency of 50 Hz. Harmonics are the multiple of the fundamental frequency, as shown in Fig. 2. Total harmonic distortion is the contribution of all the harmonic frequency currents to the fundamental.

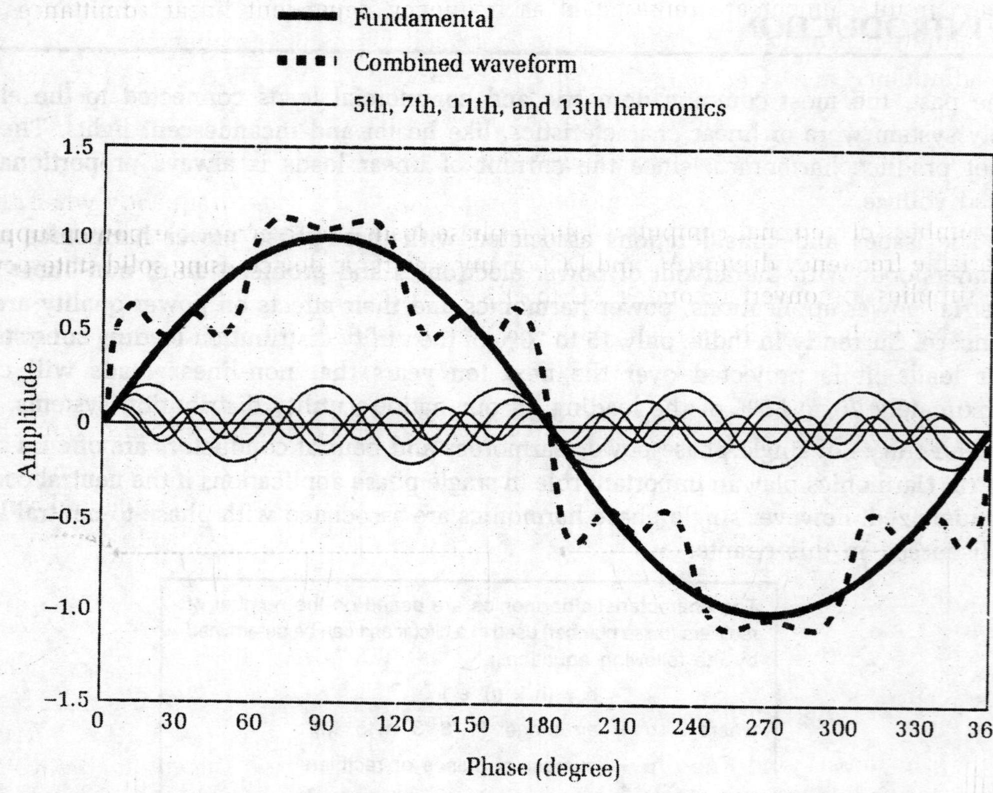

Fig. 2. *Effect of harmonics on normal voltage or current waveform.*

4.2. HARMONICS ANALYSIS

Harmonics analysis is conducted in order to study the characteristics of a power system over a range of harmonics frequencies resulting from the injection of known harmonics currents. The study will determine the magnitude of the various harmonics currents throughout the system and the voltage distortion at the various busbars.

A simulation program called Power Harmonics Simulation (PHS) was developed for this. The program first conducts a load flow study on the power system based on the Gauss–Seidel method. The bus admittance matrix and the current injection matrix at each harmonics frequency are then formulated. A nodal admittance matrix technique is employed to solve for the node voltage of each harmonics frequency.

To obtain the current through any system component, individual harmonics currents through that component are evaluated and superimposed. Each harmonics current is obtained by simply dividing the corresponding harmonics voltage across the component by its harmonics impedance.

In harmonics penetration studies, the currents from harmonics-producing elements are assumed to be known. Hence the non-linear element can be represented by a harmonics current source whose magnitude and phase angle at each frequency are fixed by the nature of devices. With the current information known, the current injection matrix for each harmonics frequency (I matrix) is easily constructed.

Linear circuit element are represented as frequency dependent linear admittance and hence the admittance matrix can be formulated for each harmonics frequency.

The admittance matrix equation is :

$$[I_n] = [Y_n]\ [V_n]$$

The unknown $[V_n]$ is solved using standard techniques.

Harmonics are the by-products of modern electronics. They occur frequently when there are large numbers of personal computers (single phase loads), uninterruptible power supplies (UPSs), variable frequency drives (AC and DC) or any electronic device using solid state power switching supplies to convert incoming AC to DC.

Non-linear loads create harmonics by drawing current in abrupt short pulses, rather than in a smooth sinusoidal manner (see Fig. 3).

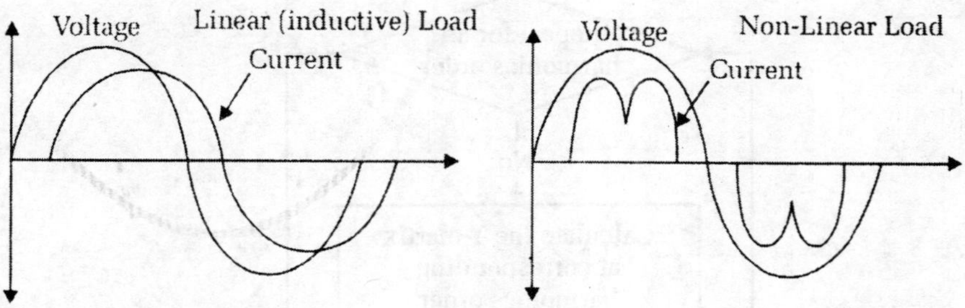

Fig. 3. *Differences between linear and non-linear loads.*

The terms "linear" and "non-linear" define the relationship of current to the voltage waveform. A linear relationship exists between the voltage and current, which is typical of an across-the-line load. A non-linear load has a discontinuous current relationship that does not correspond to the applied voltage waveform.

4.2.1. Load Flow Studies

Load flow study means the determination of voltage magnitude, $|V|$, phase angle, d, real power injection, P, and reactive power injection, Q, at each node of a power system. There are a number of analysis methods available for load flow studies, some of which are the Gauss–Seidel method, the Newton–Raphson method, and the Fast–Decoupled method. The Gauss–Seidel method is used in this paper to solve for the fundamental component of the power system.

4.2.1.1. Gauss–Seidel Method

The Gauss–Seidel (GS) method is an iterative algorithm for solving a set of nonlinear algebraic equations. To start with, a solution vector is assumed, based on guidance from practical experience in a physical situation. One of the equations is then used to obtain the revised value of a particular variable by substituting in it the present values of the remaining variables. The solution vector is immediately updated in respect of this variable. The process is then repeated for all the variables, thereby completing one-iteration. The iteration process is then repeated till the solution vector converges within prescribed accuracy. Flow chart of this method is shown in Fig. 4.

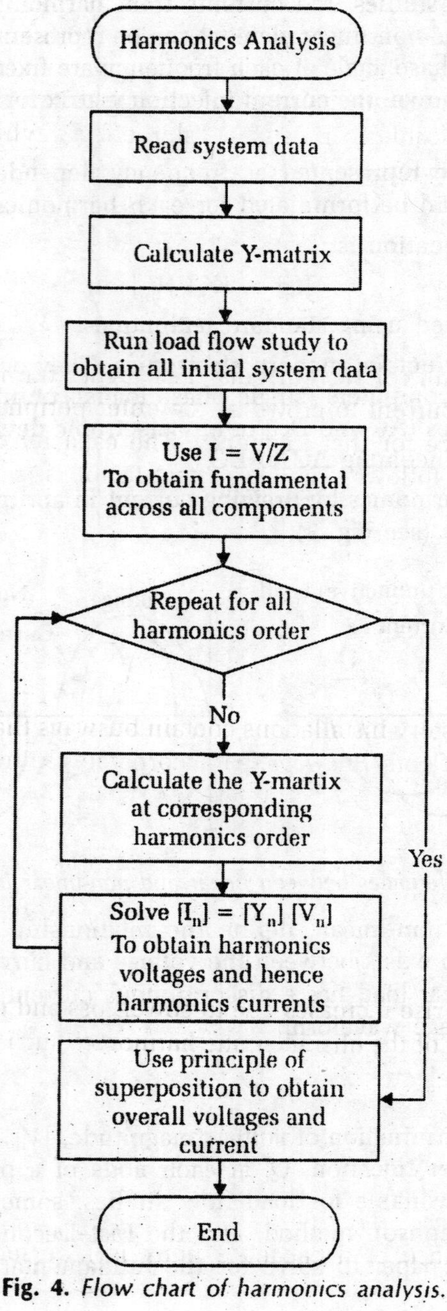

Fig. 4. *Flow chart of harmonics analysis.*

4.3. EFFECTS OF HARMONICS ON POWER SYSTEM DEVICES

The effects of three-phase harmonics on circuits are similar to the effects of stress and high blood pressure on the human body. High levels of stress or harmonic distortion can lead to problems for the utility's distribution system, plant distribution system and any other equipment serviced by that distribution system. Effects can range from spurious operation of equipment to a shutdown of important plant equipment, such as machines or assembly lines. Harmonics can lead to power system inefficiency. Some of the negative ways that harmonics may affect plant equipment are listed below :

4.3.1. Conductor/Busways Overheating

It is a function of the square rms current per unit volume of the conductor. Harmonic currents on undersized conductors or cables can cause a "skin effect", which increases with frequency and is similar to a centrifugal force.

Current flowing in a cable produces I^2R losses. When the load current contains harmonic content, additional losses are introduced. To compound the problem, the effective resistance of the cable increases with frequency because of the phenomenon known as skin effect. Skin effect is due to unequal flux linkage across the cross section of the conductor which causes AC currents to flow only on the outer periphery of the conductor. This has the effect of increasing the resistance of the conductor for AC currents. The higher the frequency of the current, the greater the tendency of the current to crowd at the outer periphery of the conductor and the greater the effective resistance for that frequency. The capacity of a cable to carry nonlinear loads may be determined as follows. The skin effect factor is calculated first. The skin effect factor depends on the skin depth, which is an indicator of the penetration of the current in a conductor. Skin depth (δ) is inversely proportional to the square root of the frequency :

$$\delta = S \div \sqrt{f}$$

Where S is proportionality constant based on the physical characteristics of the cable and its magnetic permeability and f is the frequency of the current.

Most commercial multistory installations contain busways that serve as the primary source of electrical power to various floors. Busways that incorporate sandwiched busbars are susceptible to nonlinear loading, especially if the neutral bus carries large levels of triplen harmonic currents (third, ninth, etc.). Under the worst possible conditions, the neutral bus may be forced to carry a current equal to 173% of the phase currents. In cases where substantial neutral currents are expected, the busways must be suitably derated.

4.3.2. Capacitors

They can be affected by heat rise increases due to power loss and reduced life on the capacitors If a capacitor is tuned to one of the characteristic harmonics such as the 5th or 7th, overvoltage and resonance can cause dielectric failure or rupture the capacitor.

Capacitor banks are commonly found in commercial and industrial power systems to correct for low power factor conditions. Capacitor banks are designed to operate at a maximum voltage of 110% of their rated voltages and at 135% of their rated kVARs. When large levels of voltage and current harmonics are present, the ratings are quite often exceeded, resulting in failures. Because the reactance of a capacitor bank is inversely proportional to frequency, harmonic currents can find their way into a capacitor bank. The capacitor bank acts as a sink absorbing stray harmonic currents and causing overloads and subsequent failure of the bank

A more serious condition with potential for substantial damage occurs due to a phenomenon called harmonic resonance. Resonance conditions are created when the inductive and capacitive reactances become equal at one of the harmonic frequencies. The two types of resonances are series and parallel. In general, series resonance produces voltage amplification and parallel resonance results in current multiplication. Resonance will not be analyzed in this book, but many textbooks on electrical circuit theory are available that can be consulted for further explanation. In a harmonic-rich environment, both series and parallel resonance may be present. If a high level of harmonic voltage or current corresponding to the resonance frequency exists in a power system, considerable damage to the capacitor bank as well as other power system devices can result.

4.3.3. Fuses and Circuit Breakers

Harmonics can cause false or spurious operations and trips, damaging or blowing components for no apparent reason. Harmonic currents influence the operation of protective devices. Fuses and motor thermal overload devices are prone to nuisance operation when subjected to nonlinear currents. This factor should be given due consideration when sizing protective devices for use in a harmonic environment. Electromechanical relays are also affected by harmonics. Depending on the design, an electromechanical relay may operate faster or slower than the expected times for operation at the fundamental frequency alone. Such factors should be carefully considered prior to placing the relays in service.

4.3.4. Transformers

There has been increase in iron and copper losses or eddy currents due to stray flux losses. This causes excessive overheating in the transformer windings. Typically, the uses of appropriate "*K* factor" rated units are recommended for non-linear loads.

Harmonics can affect transformers primarily in two ways. Voltage harmonics produce additional losses in the transformer core as the higher frequency harmonic voltages set up hysteresis loops, which superimpose on the fundamental loop. Each loop represents higher magnetization power requirements and higher core losses. A second and a more serious effect of harmonics is due to harmonic frequency currents in the transformer windings. The harmonic currents increase the net RMS current flowing in the transformer windings which results in additional I^2R losses. Winding eddy current losses are also increased. Winding eddy currents are circulating currents induced in the conductors by the leakage magnetic flux. Eddy current concentrations are higher at the ends of the windings due to the crowding effect of the leakage magnetic field at the coil extremities. The winding eddy current losses increase as the square of the harmonic current and the square of the frequency of the current. Thus, the eddy loss (*EC*) is proportional to $I^2{}_h X\ h^2$, where I^2 is the RMS value of the harmonic current of order *h*, and *h* is the harmonic frequency order or number. Eddy currents due to harmonics can significantly increase the transformer winding temperature. Transformers that are required to supply large nonlinear loads must be derated to handle the harmonics. This derating factor is based on the percentage of the harmonic currents in the load and the rated winding eddy current losses.

4.3.5. Generators/Motors

They have similar problems to transformers. Sizing and coordination is critical to the operation of the voltage regulator and controls. Excessive harmonic voltage distortion will cause multiple zero crossings of the current waveform. Multiple zero crossings affect the timing of the voltage regulator, causing interference and operation instability.

Application of distorted voltage to a motor results in additional losses in the magnetic core of the motor. Hysteresis and eddy current losses in the core increase as higher frequency harmonic voltages are impressed on the motor windings. Hysteresis losses increase with frequency and eddy current losses increase as the square of the frequency. Also, harmonic currents produce additional I^2R losses in the motor windings which must be accounted for.

Another effect, and perhaps a more serious one, is torsional oscillations due to harmonics. Two of the more prominent harmonics found in a typical power system are the fifth and seventh harmonics. The fifth harmonic is a negative sequence harmonic, and the resulting magnetic field revolves in a direction opposite to that of the fundamental field at a speed five times the fundamental. The seventh harmonic is a positive sequence harmonic with a resulting magnetic field revolving in the same direction as the fundamental field at a speed seven times the fundamental. The net effect is a magnetic field that revolves at a relative speed of six times the speed of the rotor. This induces currents in the rotor bars at a frequency of six times the fundamental frequency. The resulting interaction between the magnetic fields and the rotor-induced currents produces torsional oscillations of the motor shaft. If the frequency of the oscillation coincides with the natural frequency of the motor rotating members, severe damage to the motor can result. Excessive vibration and noise in a motor operating in a harmonic environment should be investigated to prevent failures.

Motors intended for operation in a severe harmonic environment must be specially designed for the application. Motor manufacturers provide motors for operation with ASD units. If the harmonic levels become excessive, filters may be applied at the motor terminals to keep the harmonic currents from the motor windings. Large motors supplied from ASDs are usually provided with harmonic filters to prevent motor damage due to harmonics.

4.3.6. Utility Meters

They may record measurements incorrectly, resulting in higher billings to consumers.

4.3.7. Drives/Power Supplies

They can be affected by misoperation due to multiple zero crossings. Harmonics can cause failure of the commutation circuits, found in DC drives and AC drives with silicon controlled rectifiers (SCRs).

4.3.8. Computers/Telephones

They may experience interference or failures. Some more effect can be seen in Table 1.

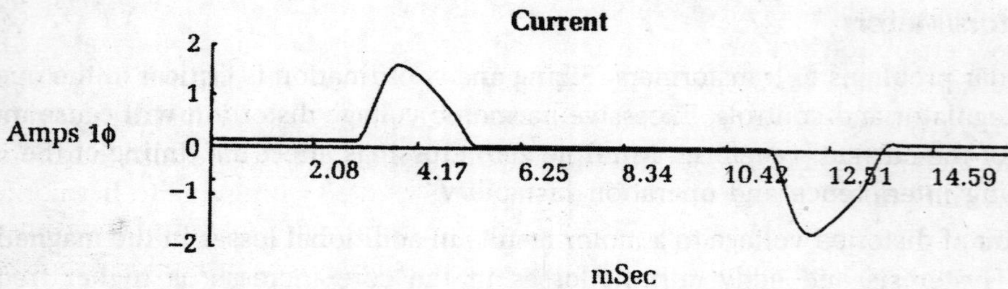

Fig. 5. *Nonlinear current drawn by single personal computer.*

Table 1 : Some Effects of Harmonics

At	*Effect*
Circuit breakers	Malfunction
Capacitor banks	Overheating
	Insulation breakdown
	Failure of internal fuses
Protection equipment	False tripping
	No tripping
Measuring devices	Wrong measurements
Transformers, reactors	Overheating
Motors	Increased noise level
	Overheating
	Additional vibrations
Telephones	Noise with the respective harmonic frequency
Lines	Overheating
Electronic devices	Wrong pulses on data transmission
	Over-/undervoltage
	Flickering screens
Incandescent lamps	Reduced lifetime
	Flicker

4.4. HARMONIC CURRENT MITIGATION

The best way to deal with harmonics problems is through prevention : choosing equipment and installation practices that minimize the level of harmonics in any one circuit or portion of a facility. Many power quality problems, including those resulting from harmonics, occur when new equipment is haphazardly added to older systems. However, even within existing facilities, the problems can often be solved with simple solutions such as fixing poor or nonexistent grounding on individual equipment or the facility as a whole, moving a few loads between branch circuits, or adding additional circuits to help isolate the sensitive equipment from what is causing the harmonic distortion.

If the problems cannot be solved by these simple measures, there are two basic choices : to reinforce the distribution system to withstand the harmonics or to install devices to attenuate or remove the harmonics. Reinforcing the distribution system means installing double-size neutral wires or installing separate neutral wires for each phase, and/or installing oversized or K rated transformers, which allow for more heat dissipation. There are also harmonic-rated circuit breakers and panels, which are designed to prevent overheating due to harmonics. This option is generally more suited to new facilities, because the costs of retrofitting an existing facility in this way could be significant. Strategies for attenuating harmonics, from cheap to more expensive, include passive harmonic filters, isolation transformers, harmonic mitigating transformers (HMTs), the Harmonic Suppression System (HSS) from Harmonics Ltd., and active filters.

Table 2 : Solutions to Harmonics Problems

Solution	***Best applications***
Reinforce distribution system (add double-sized neutral wires, harmonics-rated circuit breakers, oversized or K-rated transformer, etc.)	New facilities
Passive filter	For circuits that include three-phase loads, where there are only minor voltage imbalances between phases
Isolation transformer	Where sources of harmonics are on separate branches from harmonics-sensitive equipment
Harmonic mitigating transformer	On a moderately to heavily loaded transformer with high harmonic content in system, and/or where critical loads are backed up by an uninterruptible power supply
Harmonic suppression system	For circuits with only single-phase loads and existing or new facilities with high reliability needs, where problems exist downstream of distributiodn panel
Active harmonic filter	For circuits that include three-phase loads; voltage imbalances between phases can be present

4.4.1. Passive Filters

They (also called traps) include devices that provide low-impedance paths to divert harmonics to ground and devices that create a higher-impedance path to discourage the flow of harmonics. Both of these devices, by necessity, change the impedance characteristics of the circuits into which they are inserted. Another weakness of passive harmonic technologies is that, as their name implies, they cannot adapt to changes in the electrical systems in which they operate. This means that changes to the electrical system (for example, the addition or removal of power factor–correction capacitors or the addition of more nonlinear loads) could cause them to be overloaded or to create "resonances" that could actually amplify, rather than diminish, harmonics.

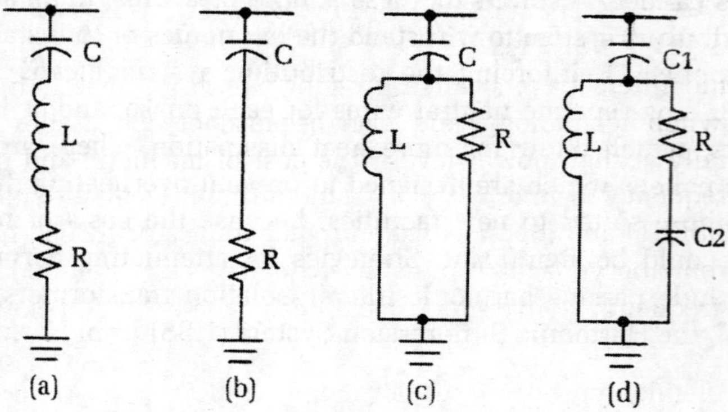

(a) Single-tuned filter
(b) First order high-pass filter
(c) Second order high-pass filter
(d) Third order high-pass filter

Fig. 6. *Passive shunt filters.*

4.4.1.1. Practical Considerations in the Use of Passive Filters

Passive power harmonic filter design requires a number of practical considerations. There is no unique solution to the design problem, so in each case a careful trade-off analysis must be performed. Practical considerations include the following :

- **Tuning** : The harmonic filter sections are tuned below the harmonic frequency to prevent the filter frequency from shifting upward if one or more capacitors fail and their fuses blow. Typical orders are 4.85 for the 5th harmonic; 6.7 for the 7th; and 10.6 for the 11th harmonic.
- **Protection** : Capacitors are protected by fuses in small groups to minimize the effect of fuse blowing. The whole filter can be divided into assemblies, each protected by a circuit breaker.
- **Switching** : Filters provide fundamental frequency reactive power (vars). Portions of the filter can be switched off at times of light load to limit overvoltage.
- **Tolerances** : Capacitors and inductors must be specified so that the combination of ratings (*L* and *C*) does not result in resonance at an undesired frequency. In other words, we do not want positive peaks in the filter impedance curves to correspond with harmonic frequencies.
- **Rating** : The current rating of the inductors and the voltage rating of the capacitors must include the fundamental and harmonic components.
- **Location** : Filters should be located electrically close to the nonlinear load that produces the harmonic currents.
- **Detuning** : A change in system impedance or component variations due to aging or temperature can result in some detuning of the harmonic filter.

4.4.2. Active Harmonic Filters

They continuously adjust their behavior in response to the harmonic current content of the monitored circuit, and they will not cause resonance. Like an automatic transmission in a car active filters are designed to accommodate a full range of expected operating conditions upon installation, without requiring further adjustments by the operator.

Active harmonic filters have been a growing area of research in recent years, due to improvements in switch technology, and also because of cost issues associated with filter components. Especially at high power levels, the cost of magnetic and capacitive components can be high. High frequency switching devices, including the metal-oxide semiconductor field-effect transistor (MOSFET) and insulated gate bipolar transistor (IGBT) have emerged in recent years with high current and voltage ratings. These devices switch on and off with fast switching speeds. Thus, high-frequency converters can be designed with good power delivery efficiency using them.

The diagram of one type of active harmonic filter is shown in Fig. 7. Note that the nonlinear load draws harmonic current, I_h, from the power source. The active compensator senses the harmonic current and injects a compensation current, I_C, which cancels the harmonic current. The net supply current, I_S, contains only the fundamental. The compensator switches at a very high frequency compared to the fundamental frequency— hence, the VA rating of the energy storage devices in the compensator can be minimized.

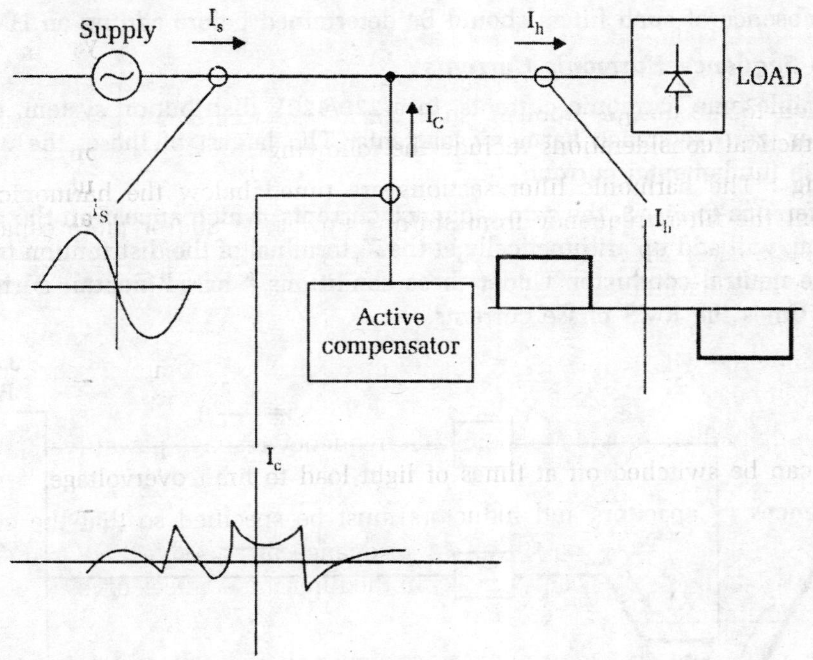

Fig. 7. *A typical active harmonic filter installation.*

4.4.2.1. Advantages of Active Filters

- **Superior filtering performance** : The active filters are generally under microprocessor control and hence can be tuned to a particular application. The filter can be tuned under microprocessor control if, for instance, the system impedance changes.
- **Smaller physical size** : The high switching speed of devices allows energy storage elements (capacitors and inductors) to be of smaller weight and volume.
- **Flexibility** : They are more flexible in application compared to passive filters. Of course, the purported advantages of active filters must be weighed against extra design time and cost.

4.4.3. Isolation Transformers

These are filtering devices that segregate harmonics in the circuit in which they are created, protecting upstream equipment from the effects of harmonics. These transformers do not remove the problem in the circuit generating the harmonics, but they can prevent the harmonics from affecting more sensitive equipment elsewhere within the facility.

4.4.4. Harmonic Mitigating Transformers

Harmonic mitigating transformers actually do relieve problematic harmonics. HMTs can be quite cost-effective in the right application, because they can both improve reliability and reduce energy costs. The right application includes transformers that are heavily or moderately loaded and where high levels of harmonic currents are present. In addition, HMTs are very effective in supporting critical loads that are backed up by a UPS. UPSs and backup generators tend to have high impedance, which results in high voltage distortion under nonlinear loading. Because of this, equipment that operates flawlessly when supplied by utility power may malfunction when the backup system engages during a utility outage. Note that some of these power systems have output filters (either passive or active) to control harmonic levels. The presence or absence of such filters should be determined before adding an HMT.

4.4.4.1. Zero Sequence Harmonic Currents

The most troublesome harmonic currents, in a 220/420V distribution system, are the 'triplin' or 'third order' zero sequence harmonic currents. The largest of these, the 3rd, is typically 60–70% of the fundamental current.

With reference to Fig. 8, the zero sequence currents, which appear on the three phases of the sub-system, will add up arithmetically at the X_0 terminal of the distribution transformer and appear on the neutral conductor. Under these conditions, 'shared' neutral currents can be as high as 1.73 times the RMS phase currents.

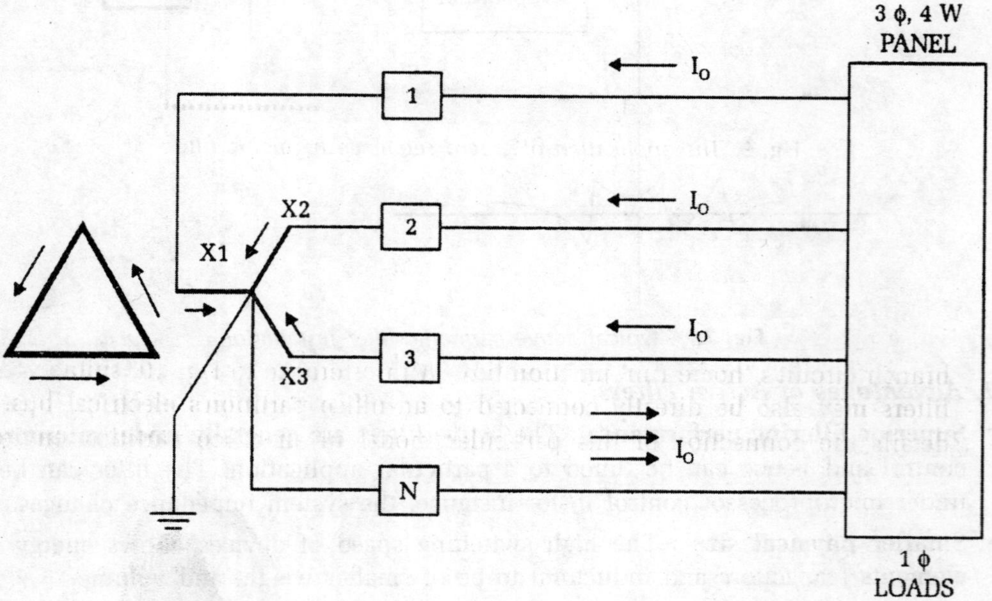

Fig. 8. *Zero sequence harmonic currents in a sub-system.*

High neutral currents, whether on 'shared' neutrals or individual neutrals, are the cause of neutral to ground voltages at the loads. With the neutral grounded at or near the distribution transformer's X_0 terminal, neutral current will generate higher levels of neutral to ground voltages, at the loads, as the circuit length (Z_0) increases ($E_0 = I_0 * Z_0$). ITIC recommends a 5V limit at the loads.

Shunting zero sequence harmonic currents from the sub-system at or near the harmonic-generating loads is the best technical solution. However, the objective is to meet IEEE and ITIC recommendations at the lowest possible cost.

There are normally four potential sites where zero sequence currents may be shunted from the sub-system. The following is a guide :

1. For low capacity sub-systems (75kVA or less), with relatively short secondary circuit lengths, ultra-low zero sequence impedance harmonic mitigating transformers will normally provide a satisfactory solution.
2. As an alternative to Item 1 and with reference to Fig. 9, a single large zero sequence harmonic shunt filter may be applied at the secondary panel of an existing sub-system without replacing the distribution transformer.

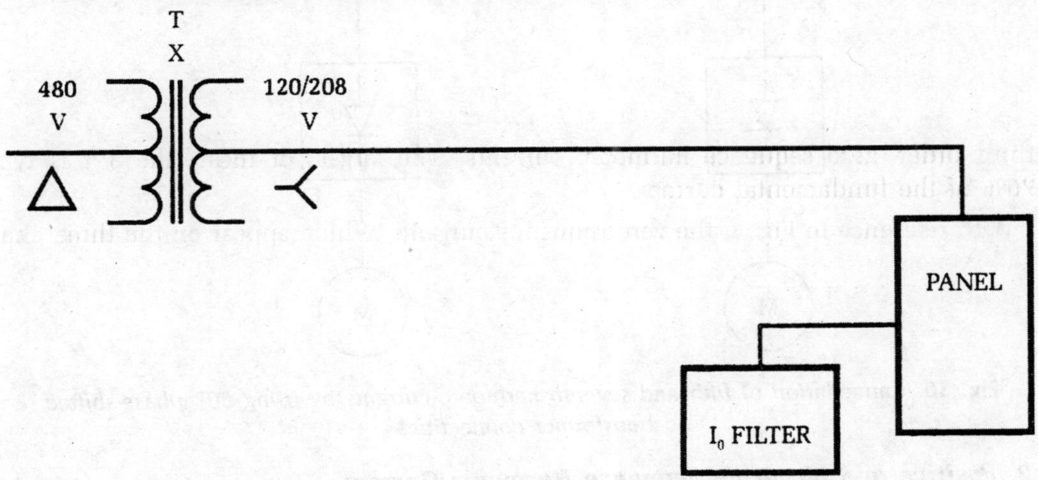

Fig. 9. *The application of a zero sequence harmonic filter.*

3. For medium capacity sub-systems, with 120/208V feeder circuits and sub-panels, there may be no other alternative than the application of smaller filters at each sub-panel.
4. For high capacity sub-systems, where the non-linear loads are remote and/or for sensitive computers that are interconnected by a communications 'network', there may be no other alternative than to apply small zero sequence harmonic shunt filters at each branch circuit's 'home run' junction box. With reference to Fig. 10, suitably packaged filters may also be directly connected to an office partition's electrical bus. Fig. 10 details the connection of this particular model in an office partition environment. 'Rack-mount' filters are also available for server and audio/video applications. In addition to shunting zero sequence harmonic currents, these devises also function as phase current balancers at the fundamental and all positive and negative sequence harmonic frequencies. This characteristic ensures minimal neutral current and neutral to ground voltages. Zero sequence filters are passive electro-magnetic devices that have none of

the negative characteristics of tuned L-C filters. In the absence of 'triplin' harmonics, they simply keep the system balanced.

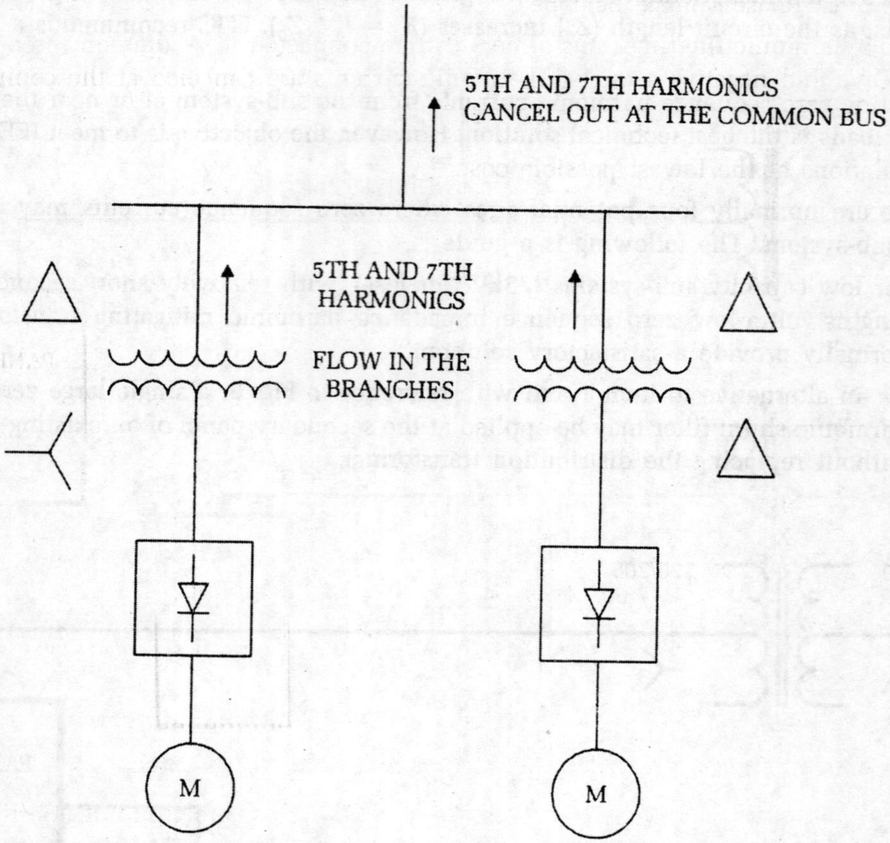

Fig. 10. *Cancellation of fifth and seventh harmonic currents by using 30° phase-shifted transformer connections.*

4.4.4.2. Positive and Negative Sequence Harmonic Currents

All non-linear electronic loads also generate positive and negative sequence harmonic currents. Cancellation of the high value, lower order positive and negative sequence harmonic currents, at a location that is as-close-as-possible to the harmonic-generating loads, is the best technical solution. Again, the objective is to meet IEEE and ITIC recommendations at the lowest possible cost.

As an alternative to the severe de-rating of conventional distribution transformers or the installation of K-Rated transformers, the application of harmonic filtering transformers, as shown in Fig. 11, will provide the following system benefits :

- Reduced Peak Phase Current
- Reduced Average Phase Current
- Reduced Total Harmonic Distortion of Current
- Reduced Total Harmonic Distortion of Voltage
- Reduced Apparatus and Circuit Losses
- Reduced Overheating

- Increased True Power Factor
- Improved Phase Current Balance
- Improved Phase Voltage Balance

Multiple harmonic filtering transformers can be connected to a common three-phase bus so that positive and negative sequence harmonic currents are canceled at the common bus.

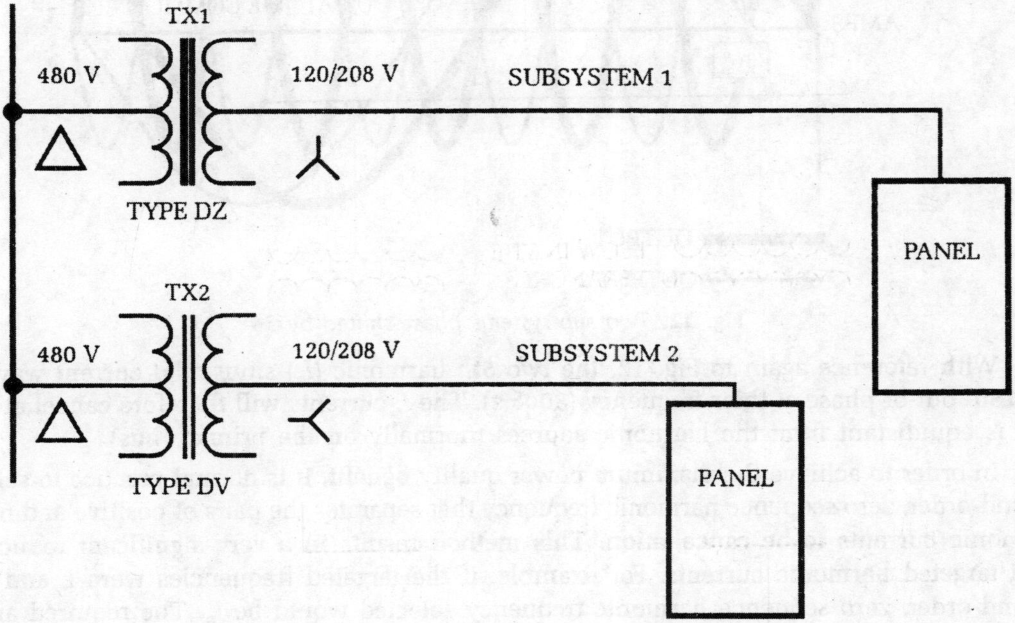

Fig. 11. *The application of harmonic filtering transformers.*

Targeted harmonic currents are canceled by phase-shifting one group of non-linear loads with reference to a second similar group of non-linear loads. The exact phase-shift angle that is required to cause perfect cancellation of a particular harmonic current, under identical loading conditions, is as follows :

$$\frac{180^\circ\ @\ 60\text{Hz}}{\text{H}} = \text{Ø}^\circ\ @\ 60\text{Hz}$$

where,

H—The harmonic number of the targeted harmonic frequency.

Ø° @ 60Hz—The phase-angle, @ 60Hz, that will cause two separate sources of the targeted harmonic current to be out-of-phase by 180° at their frequency.

The two fundamental (I_1) sinusoidal current waveforms, displayed in Fig. 12, appear at the X_1 terminals of the two harmonic filtering transformers that have a phase-shift angle of 36° between their secondary windings. The two 5th harmonic (I_5) sinusoidal current waveforms, in the same display, also appear at the X_1 terminals of same two harmonic filtering transformers.

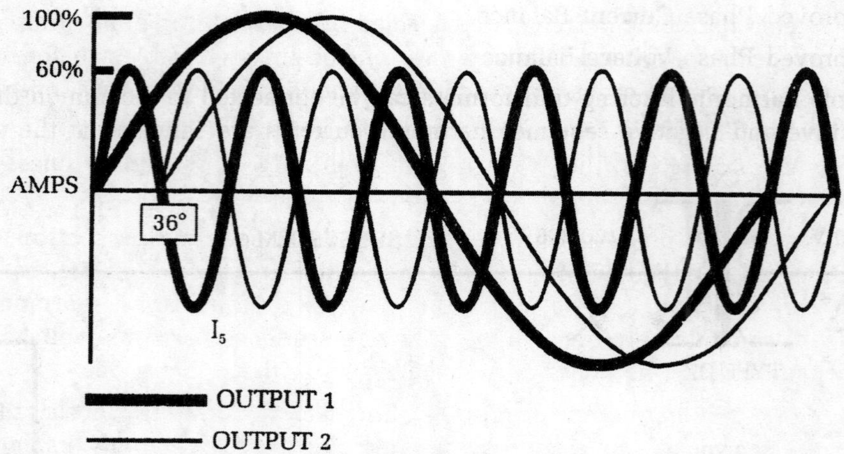

Fig. 12. *Two sub-systems phase-shifted by 36°.*

With reference again to Fig. 12, the two 5th harmonic (I_5) sinusoidal current waveforms are 180° out-of-phase at their frequency (300Hz). The I_5 currents will therefore cancel at a node that is equidistant from the harmonic sources (normally on the primary bus).

In order to achieve the maximum power quality benefit, it is normal practice to select the second-order, zero sequence harmonic frequency that separates the pairs of positive and negative harmonic currents to be cancellation. This method results in a very significant reduction of both targeted harmonic currents. For example, if the targeted frequencies were I_5 and I_7, the second-order, zero sequence harmonic frequency selected would be I_6. The required angle is, therefore, 30° (180°@ 60Hz/6 = 30°).

With reference to Fig. 11, the transformer designated as Type DZ has a primary-to-secondary phase-shift of 0° while the Type DV has a primary-to-secondary phase-shift of –30°. As shown in the Fig. 13 example, dual output—harmonic-filtering transformers can be applied so that targeted positive and negative sequence harmonic currents are canceled.

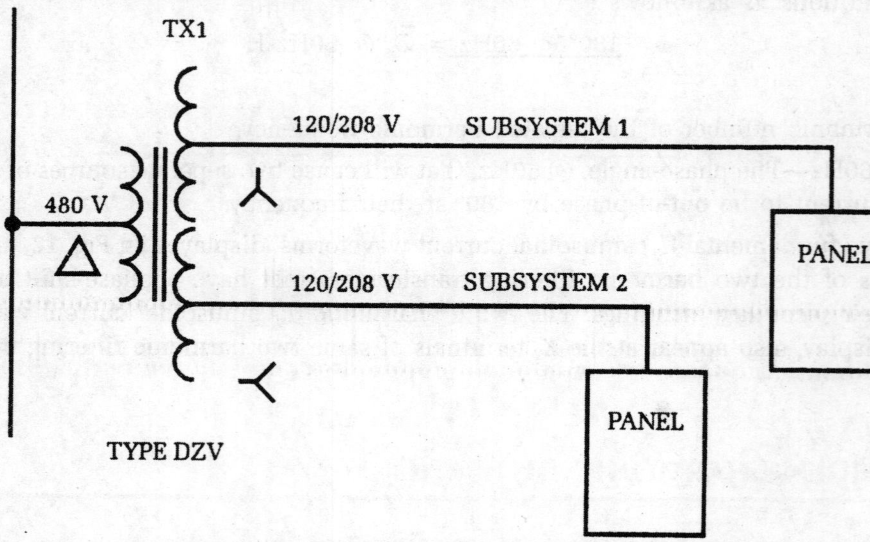

Fig. 13. *Application of dual output transformer.*

Again, targeted harmonic currents are canceled by phase-shifting one group of non-linear loads with reference to a second similar group of non-linear loads. As before, it is normal practice to select the second-order, zero sequence harmonic frequency that separates the pairs of positive and negative sequence harmonic currents to be can ellation.

With reference to Fig. 13, the Type DZV has a primary to secondary phase-shift of both 0° and –30°.

In addition to the phase-shifting techniques described in this section, all of these transformers can be supplied with zero sequence impedances of $<0.005\Omega$, as compared to $<0.1\Omega$ for a conventional or K-Rated transformer. As a result, these harmonic-mitigating transformers will generate only trace amounts of zero sequence harmonic voltage. In addition, these transformers will act as phase current balancers at the primary bus.

As a result of these harmonic current mitigating techniques, THD_V levels are significantly reduced in both the primary and secondary systems. The THD_V reductions displayed in Fig. 14 are typical when conventional and K-Rated distribution transformers are replaced with single or dual output harmonic mitigating transformers. It should be understood, however, that the transformer's phase shift is not the only important characteristic. Positive and negative harmonic impedance is critical with respect to voltage distortion at their secondary terminals.

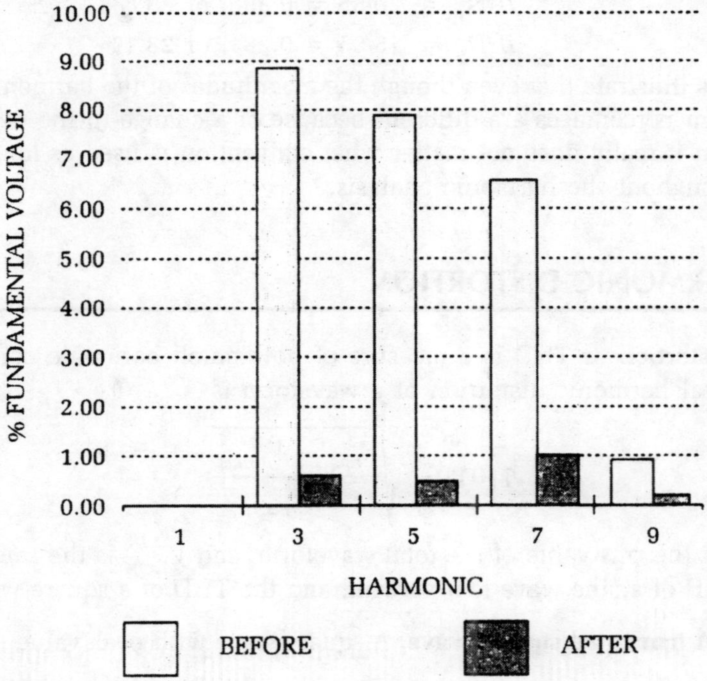

Fig. 14. *Harmonic voltage profiles for non-linear loads with harmonic filtering transformers.*

4.5. INDIVIDUAL HARMONIC DISTORTION

Individual harmonic distortion (IHD) is the ratio between the root mean square (RMS) value of the individual harmonic and the RMS value of the fundamental

$$IHD_n = I_n/I_1$$

Example 1 : *Assume that the RMS value of the third harmonic current in a nonlinear load is 20 A, the RMS value of the fifth harmonic current is 15 A, and the RMS value of the fundamental is 60 A. Then, the individual third harmonic distortion is :*

$$IHD_3 = 20/60 = 0.333, \text{ or } 33.3\%$$

And the individual fifth harmonic distortion is :

$$IHD_5 = 15/60 = 0.25, \text{ or } 25.0\%$$

Under this definition, the value of IHD_1 is always 100%. This method of quantifying the harmonics is known as harmonic distortion based on the fundamental. This is the convention used by the Institute of Electrical and Electronic Engineers (IEEE) in the U.S. The European International Electrotechnical Commission (IEC) quantifies harmonics based on the total RMS value of the waveform. Using the same example shown above, the RMS value of the waveform is :

$$I_{rms} = \sqrt{\left(60^2 + 20^2 + 15^2\right)} = 65 \text{ A}$$

Based on the IEC convention,

$$IHD_1 = 60/65 = 0.923, \text{ or } 92.3\%$$

$$IHD_3 = 20/65 = 0.308, \text{ or } 30.8\%$$

$$IHD_5 = 15/65 = 0.231, \text{ or } 23.1\%$$

The examples illustrate that even though the magnitudes of the harmonic currents are the same, the distortion percentages are different because of a change in the definition. It should be pointed out that it really does not matter what convention is used as long as the same one is maintained throughout the harmonic analysis.

4.6. TOTAL HARMONIC DISTORTION

Total harmonic distortion or THD is a measure of how much harmonic content there is in a waveform. The total harmonic distortion of a waveform is

$$\text{THD} = \sqrt{\frac{V_{rms}^2 - V_{1,\,rms}^2}{V_{1,\,rms}^2}}$$

Where V_{rms} is the rms value of the total waveform, and $V_{1,rms}$ is the *rms* value of the first harmonic. The THD of a sine wave is 0 percent, and the THD of a square wave is 48 percent.

Example 2 : A truncated square wave. A square wave with peak value +1 has the Fourier series :

$$v(t) = \sum_{n=1}^{\infty} \left(\frac{4}{n\pi}\right) \sin(2\pi nt)$$

A truncated Fourier series approaches the ideal waveform. We'll next find the rms value and total harmonic distortion for a square wave waveform with harmonics present up to the eventh. The total waveform for this example is

$$v(t) = \left(\frac{4}{\pi}\right)\sin(\omega t) + \left(\frac{4}{3\pi}\right)\sin(3\omega t) + \left(\frac{4}{5\pi}\right)\sin(5\omega t) + \left(\frac{4}{7\pi}\right)\sin(7\omega t)$$

The *rms* value of the first harmonic is

$$V_{1,rms} = \frac{4}{\pi\sqrt{2}} = 0.9$$

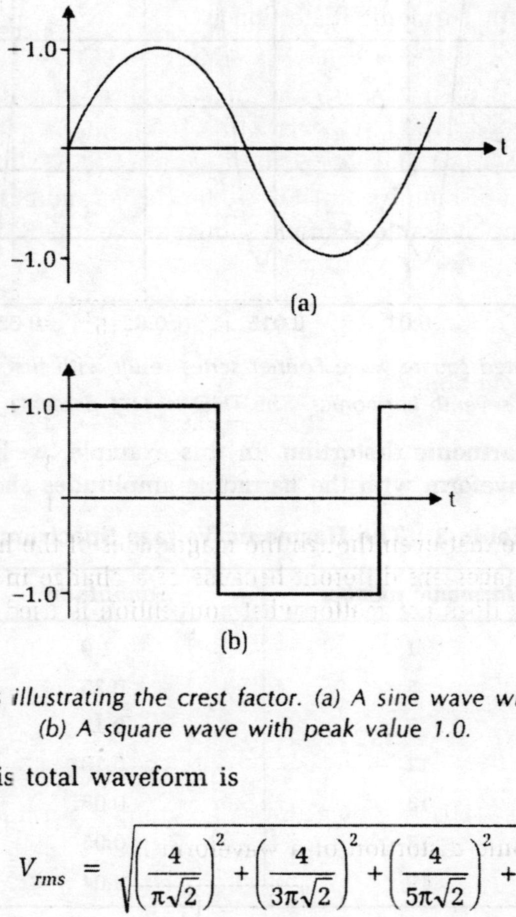

Fig. 15. *Waveforms illustrating the crest factor. (a) A sine wave with peak value 1.0. (b) A square wave with peak value 1.0.*

The *rms* value of this total waveform is

$$V_{rms} = \sqrt{\left(\frac{4}{\pi\sqrt{2}}\right)^2 + \left(\frac{4}{3\pi\sqrt{2}}\right)^2 + \left(\frac{4}{5\pi\sqrt{2}}\right)^2 + \left(\frac{4}{7\pi\sqrt{2}}\right)^2}$$

$$= \left(\frac{4}{\pi\sqrt{2}}\right)\sqrt{1+\left(\frac{1}{3}\right)^2 + \left(\frac{1}{5}\right)^2 + \left(\frac{1}{7}\right)^2} = 0.974$$

The total harmonic distortion is

$$\text{THD} = \sqrt{\frac{V_{rms}^2 - V_{1,rms}^2}{V_{1,rms}^2}} = \sqrt{\frac{(0.974)^2 - (0.9)^2}{(0.9)^2}} = 0.414 = 41.4\%$$

This waveform is shown in Fig. 16.

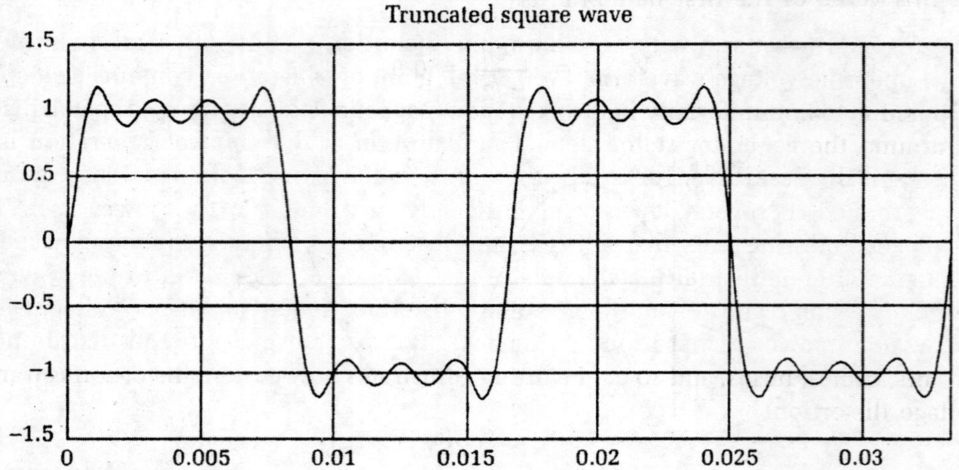

Fig. 16. *A truncated square wave Fourier series result with first, third, fifth, and seventh harmonics. The THD is 41.4 percent.*

Example 3 : Total harmonic distortion. In this example, we'll find the total harmonic distortion for a voltage waveform with the harmonic amplitudes shown in Table 3.

Table 3 : The Harmonic Voltage Spectrum

Harmonic number	*Amplitude*
1	1.0
5	0.25
7	0.15
11	0.10
13	0.08
17	0.05
19	0.04

The THD for this waveform is found simply by :

$$\text{THD} = \frac{\sqrt{0.25^2 + 0.15^2 + 0.10^2 + 0.08^2 + 0.05^2 + 0.04^2}}{1.0} = 33.5\%$$

The individual harmonic distortion indicates the contribution of each harmonic frequency to the distorted waveform, and the total harmonic distortion describes the net deviation due to all the harmonics. These are both important parameters. In order to solve harmonic problems, we require information on the composition of the individual distortions so that any treatment may be tailored to suit the problem. The total harmonic distortion, while conveying no information on the harmonic makeup, is used to describe the degree of pollution of the power system as far as harmonics are concerned. Defining the individual and total harmonic distortions will be helpful as we look at some typical nonlinear waveforms and their harmonic frequency characteristics.

4.7. CAUSES OF VOLTAGE AND CURRENT HARMONICS

A pure sinusoidal waveform with zero harmonic distortion is a hypothetical quantity and not a practical one. The voltage waveform, even at the point of generation, contains a small amount of distortion due to nonuniformity in the excitation magnetic field and discrete spatial distribution of coils around the generator stator slots. The distortion at the point of generation is usually very low, typically less than 1.0%. The generated voltage is transmitted many hundreds of miles, transformed to several levels, and ultimately distributed to the power user. The user equipment generates currents that are rich in harmonic frequency components, especially in large commercial or industrial installations. As harmonic currents travel to the power source, the current distortion results in additional voltage distortion due to impedance voltages associated with the various power distribution equipment, Such as transmission and distribution lines, transformers, cables, buses, and so on. Figure 17, illustrates how current distortion is transformed into voltage distortion.

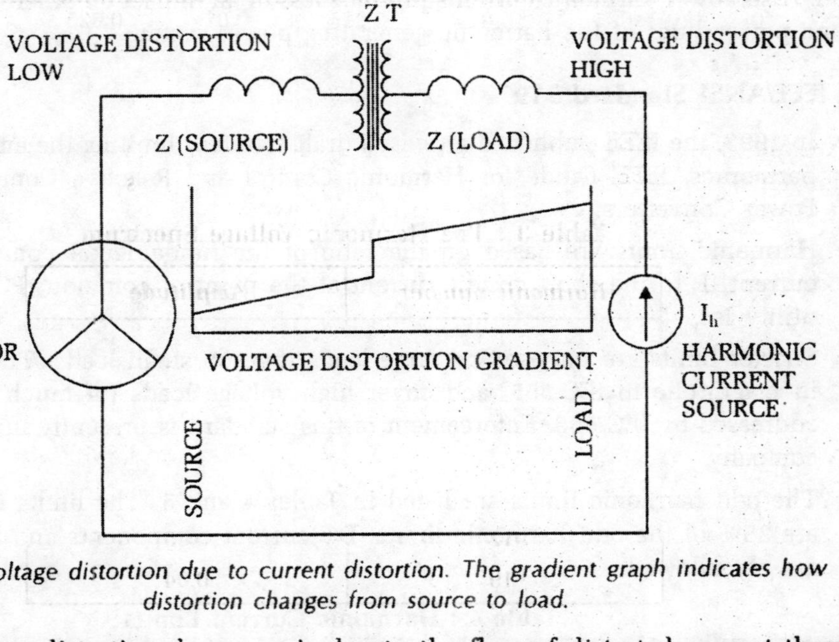

Fig. 17. *Voltage distortion due to current distortion. The gradient graph indicates how distortion changes from source to load.*

Not all voltage distortion, however, is due to the flow of distorted current through the power system impedance. For instance, static uninterruptible power source (UPS) systems can generate appreciable voltage distortion due to the nature of their operation. Normal AC voltage is converted to DC and then reconverted to AC in the inverter section of the UPS. Unless waveform shaping circuitry is provided, the voltage waveforms generated in UPS units tend to be distorted.

As nonlinear loads are propagated into the power system, voltage distortions are introduced which become greater moving from the source to the load because of the circuit impedances. Current distortions for the most part are caused by loads. Even loads that are linear will generate nonlinear currents if the supply voltage waveform is significantly distorted. When several power users share a common power line, the voltage distortion produced by harmonic current injection of one user can affect the other users. This is why standards are being issued that will limit the amount of harmonic currents that individual power users can feed into the source (an issue that we will examine later in this chapter). The major causes of current distortion are nonlinear loads due to adjustable speed drives, fluorescent lighting, rectifier banks, computer and data-processing loads, arc furnaces, and so on. One can easily visualize an environment where a wide spectrum of harmonic frequencies are generated and transmitted to other loads or other power users, thereby producing undesirable results throughout the system.

4.8. GUIDELINES FOR HARMONIC VOLTAGE AND CURRENT LIMITATION

So far we have discussed the adverse effects of harmonics on power system operation. It is important, therefore, that attempts be made to limit the harmonic distortion that a facility might produce. There are two reasons for this. First, the lower the harmonic currents produced in an electrical system, the better the equipment within the confinement of the system will perform. Also, lower harmonic currents produce less of an impact on other power users sharing the same power lines of the harmonic generating power system.

4.8.1. IEEE/ANSI Standard 519

- In 1993, the IEEE published a revised draft standard limiting the amplitudes of current harmonics, IEEE Guide for Harmonic Control and Reactive Compensation of Static Power Converters.
- Harmonic limits are based on the ratio of the fundamental component of the load current IL to the short circuit current at the point of common (PCC) coupling at the utility Isc.
- Stricter limits are imposed on large loads than on small loads. The limits are similar in magnitude to IEC-555, and cover high voltage loads (of much higher power) not addressed by IEC-555. Enforcement of this standard is presently up to the local utility company.
- The odd harmonic limits are listed in Tables 4 and 5. The limits for even harmonics are 25% of the odd harmonic limits. DC current components and half-wave rectifiers are not allowed.

Table 4 : Harmonic Current Limits

I_{SC}/I_L	$h < 11$	$11 \le h < 17$	$17 \le h < 23$	$23 \le h < 35$	$35 \le h$	*THD*
< 20	4.0	2.0	1.5	0.6	0.3	5.0
20–50	7.0	3.5	2.5	1.0	0.5	8.0
50–100	10.0	4.5	4.0	1.5	0.7	12.0
100–1000	12.0	5.5	5.0	2.0	1.0	15.0
> 1000	15.0	7.0	6.0	2.5	1.4	20.0

Note : I_{SC} = maximum short-circuit current at PCC; I_L = maximum fundamental frequency demand load current at PCC (average current of the maximum demand for the preceding 12 months); h = individual harmonic order; THD = total harmonic distortion based on the maximum demand load current. The table applies to odd harmonics; even harmonics are limited to 25% of the odd harmonic limits shown above.

As the ratio between the maximum available short circuit current at the PCC and the maximum demand load current increases, the percentage of the harmonic currents that are allowed also increases. This means that larger power users are allowed to inject into the system only a minimal amount of harmonic current (as a percentage of the fundamental current). Such a scheme tends to equalize the amounts of harmonic currents that large and small users of power are allowed to inject into the power system at the PCC. IEEE 519 also provides guidelines for maximum voltage distortion at the PCC (see Table 5). Limiting the voltage distortion at the PCC is the concern of the utility. It can be expected that as long as a facility's harmonic current contribution is within the IEEE 519 limits the voltage distortion at the PCC will also be within the specified limits.

Table 5 : Voltage Harmonic Distortion Limits

Bus Voltage at PCC	*Individual Voltage Distortion (%)*	*Total Voltage Distortion THD (T)*
69 kV and below	3.0	5.0
69.001 kV through 161 kV	1.5	2.5
161.001 kV and above	1.0	1.5

Note : PCC = point of common coupling; THD = total harmonic distortion.

When the IEEE 519 harmonic limits are used as guidelines within a facility, the PCC is the common junction between the harmonic generating loads and other electrical equipment in the power system. It is expected that applying IEEE guidelines renders power system operation more reliable. In the future, more and more utilities might require facilities to limit their harmonic current injection to levels stipulated by IEEE 519. The following section contains information on how harmonic mitigation might be achieved.

4.8.2. International Electrotechnical Commission Standard 555

- First draft of their IEC-555 standard :1982. It has since undergone a number of revisions.
- Enforcement of IEC-555 is the prerogative of each individual country, and hence it has been sometimes difficult to predict whether and where this standard will actually be given the force of law.
- Nonetheless, IEC-555 is now enforced in Europe, making it a de facto standard for commercial equipment intended to be sold worldwide.
- IEC-555 covers a number of different types of low power equipment, with differing harmonic limits. Harmonics for equipment having an input current of up to 16A, connected to 50 or 60 Hz, 220V to 240V single phase circuits (two or three wire), as well as 380V to 415V three phase (three or four wire) circuits, are limited.

4.8.2.1. IEC-555 : Class A and B

Class A : Balanced three-phase equipment, and any equipment which does not fit into the other categories. This class includes low harmonic rectifiers for computer and other office equipment. These limits are given in Table 6, and are absolute ampere limits.

Class B : Portable tools, and similar devices. The limits are equal to the Table 6 limits, multiplied by 1.5.

Table 6 : IEC-555 Harmonic Current Limits, Class A and Certain Class C

Odd harmonics		*Even harmonics*	
Harmonic number	***Maximum current***	***Harmonic number***	***Maximum current***
3	2.30A	2	1.08A
5	1.14A	4	0.43A
7	0.77A	6	0.30A
9	0.40A	$8 \le n \le 40$	$0.23\ A \cdot (8/n)$
11	0.33A		
13	0.21A		
$15 \le n \le 39$	$0.15A \cdot (15/n)$		

4.8.2.2. IEC-555 : Class C

Lighting equipment, including dimmers and gas discharge lamps. The input current harmonics of ballasted lamps with an input power of more than 25W must meet the limits of Table 7, expressed as a percent of the fundamental current. If the input power is less than 25W, then Table 8 applies. Incandescent lamp fixtures containing phasecontrol dimmers, rated at greater than 600W, must meet the limits of Table 7. When testing for compliance, the dimmer must drive a rated-power lamp, with the phase control set to a firing angle of 90°±5°. Incandescent lamp dimmers rated at less than 600W are not covered by the standard. Discharge lamps containing dimmers must meet the limits of both Tables 7 and 8 at maximum load. Harmonic currents at any dimming position may not exceed the maximum load harmonic currents.

Table 7 : IEC-555 Harmonic Current Limits, Certain Class C

Harmonic number	***Maximum current, percent of fundamental***
2	2%
3	(30%) · (power factor)
5	10%
7	7%
9	5%
$11 \le n \le 39$	3%

Table 8 : IEC-555 Harmonic Current Limits, Class D and Certain Class C

Odd harmonics			*Even harmonics*		
Harmonic number	***Relative limit (mA/W)***	***Absolute limit (A)***	***Harmonic number***	***Relative limit (mA/W)***	***Absolute limit (A)***
3	3.6	2.30A	2	1.0	0.3A
5	2.0	1.14A ·	4	0.5	0.15A
7	1.5	0.77A			
9	1.0	0.40A			
$11 \le n \le 39$	$0.6 \cdot (11/n)$	0.33A			

4.8.2.3. *IEC-555 : Class D*

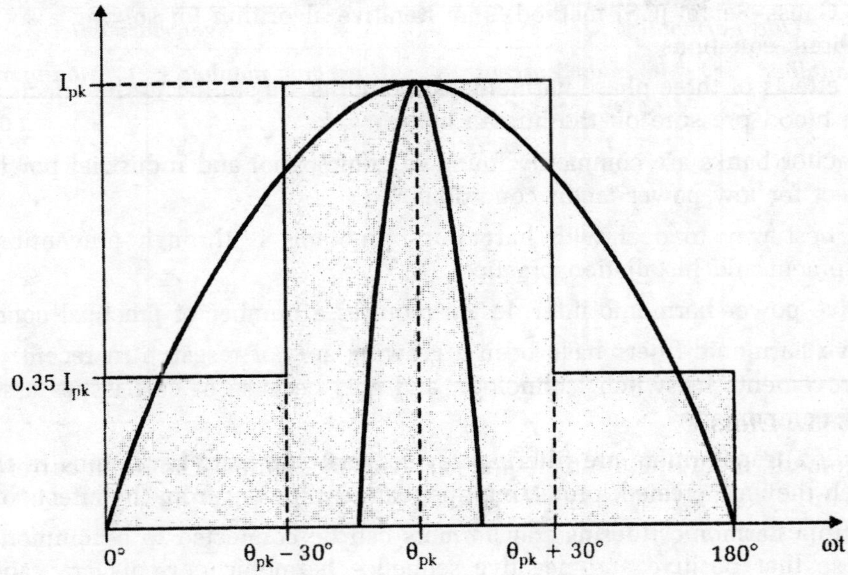

Fig. 18. *Power and harmonics in nonsinusoidal systems.*

Equipment not covered by Class *B* or *C*, which has a special waveshape as defined below. This class is directed at diode-capacitor peak-detection rectifiers. Equipment is placed in class *D* if the ac input current waveshape lies within the shaded area for at least 95% of the duration of each half-cycle. The center line of the shaded area is set to coincide with the peak of the current waveform. A sinusoid, and a typical peak detection rectifier waveform, are shown for reference; the sinusoid is not Class *D* but the peak detection rectifier waveform is the limits for Class *D* equipment are given in Table 8.

4.8.3. Low-Power Harmonic Limits

- In a city environment such as a large building, a large fraction of the total power system load can be nonlinear.
- Example : a majcr portion of the electrical load in a building is comprised of fluorescent lights, which present a very nonlinear characteristic to the utility system.
- A modern office may also contain a large number of personal computers, printers, copiers, etc., each of which may employ peak detection rectifiers.
- Although each individual load is a negligible fraction of the total local load, these loads can collectively become significant.

SUMMARY

- A harmonic is a component of a periodic wave having a frequency that is an integral multiple of the fundamental power line frequency of 50 Hz.
- Harmonics analysis is conducted in order to study the characteristics of a power system over a range of harmonics frequencies resulting from the injection of known harmonics currents.

- Harmonics are the by-products of modern electronics.
- The Gauss–Seidel (GS) method is an iterative algorithm for solving a set of nonlinear algebraic equations.
- The effects of three-phase harmonics on circuits are similar to the effects of stress and high blood pressure on the human body.
- Capacitor banks are commonly found in commercial and industrial power systems to correct for low power factor conditions.
- The best way to deal with harmonics problems is through prevention : choosing equipment and installation practices.
- Passive power harmonic filter design requires a number of practical considerations.
- Active harmonic filters have been a growing area of research in recent years, due to improvements in switch technology, and also because of cost issues associated with filter components.
- Isolation transformers are filtering devices that segregate harmonics in the circuit in which they are created, protecting upstream equipment from the effects of harmonics.
- Multiple harmonic filtering transformers can be connected to a common three-phase bus so that positive and negative sequence harmonic currents are canceled at the common bus.
- In order to achieve the maximum power quality benefit, it is normal practice to select the second-order, zero sequence harmonic frequency that separates the the pairs of positive and negative harmonic currents to be cancellation.
- Harmonic limits are based on the ratio of the fundamental component of the load current.
- The individual harmonic distortion indicates the contribution of each harmonic frequency to the distorted waveform.
- The total harmonic distortion describes the net deviation due to all the harmonics.

Review Questions and Exercises

1. Define harmonics. Explain the causes of current harmonics. *(MDU, Rohtak, May 2008)*
2. Explain harmonic current mitigation. *(MDU, Rohtak, May 2008, May 2009)*
3. Explain ill effects of harmonics on the performance of power system devices. *(MDU, Rohtak, May 2008)*
4. Explain the guidelines for harmonic voltage and current limitation. *(MDU, Rohtak, Dec. 2008, May 2009)*
5. Explain harmonic current mitigation by active filters. *(MDU, Rohtak, Dec. 2008)*
6. Explain the terms individual and total harmonic distortion. *(MDU, Rohtak, Dec. 2008)*
7. Explain the harmonics in D.C. voltages. *(MDU, Rohtak, Dec. 2008, May 2009)*
8. Define harmonics. Explain various types of harmonics. *(MDU, Rohtak, May 2009)*
9. Explain various causes of harmonics in a power *(MDU, Rohtak, May 2009)*

□□□

Chapter 5

MEASURING AND SOLVING POWER QUALITY PROBLEMS

5.1. INTRODUCTION

In the new context of open electricity market, system operators will have to report more and more to external parties, namely to users and to regulators, about power system performances in terms of Power Quality (PQ) issues. Indeed, a number of regulators have already defined, or are planning to establish, power quality objectives (supply continuity and voltage quality) to be met by the electricity supply systems. In some countries, regulators may even impose penalties in case of non-observance of the power quality objectives. The need of a distributed monitoring of PQ indexes becomes a crucial point.

In general, with reference to PQ issues, two main aspects are focused on : Commercial Quality and Quality of Supply; the former concerns mainly legal and commercial interests, such as contracts, customer satisfaction and so on, the latter deals with more technical aspects. In this context, it is necessary to monitor both the behavior of "sensitive" customers and that of "aggressive" customers contributing to PQ degradation.

The information acquired by PQ monitoring systems, in particular with reference to conducted disturbances, must be stored and analyzed remotely from the measuring point. With the widespread diffusion of computer networks and Internet, and also thanks to the introduction of tools for distributed monitoring systems design in the market, the trend is to built up even more sophisticated systems. The measurement instruments are connected to the network and the data are transferred to a database accessible via web by engineers, managers or ancillary service operators through the local area network (LAN) and from customers through a wide area network (WAN).

Several PQ monitoring systems are available in commerce but almost all of them have the big disadvantage of being embedded and close. This means that it is very difficult to connect different instruments built by different manufacturers to the same system infrastructure and/

or to merge all the monitoring results for unified statistical analyses. Moreover, the commercial PQ monitoring systems are made up for the general and standard purpose of analyses so they do not always meet the specific customer demands.

5.2. POWER QUALITY MEASUREMENTS

The basic idea is to realize a distributed PQ monitoring system that allows measuring and characterizing the previous quantities, controlling and analyzing more points of the same and/ or of different electrical network systems. This can be obtained by distributing the processing power, utilizing both commercial and non-commercial instruments, with inherent processing capabilities, that can communicate between them using numerical, fast and reliable transmission systems. This measurement system allows verifying the functioning of each single device from a remote location by means of high-level user interfaces.

In order to satisfy this target, the proposed measurement system has been designed to implement a full industrial bus architecture structured in three hierarchical communication levels, as illustrated by Fig. 1 : Fieldbus, LAN (Local Area Network) and WAN (Wide Area Network). The fieldbus links field devices to a fieldbus server. The LAN interconnects fieldbus servers and measurement servers/clients. Remote clients on the WAN can access the LAN through a gateway. All the instruments are able to process the acquired data and to transmit all the results to the data server on the network. The field-bus server acquires the data coming from several instruments and, after local analysis; it can transmit the result to the generic client.

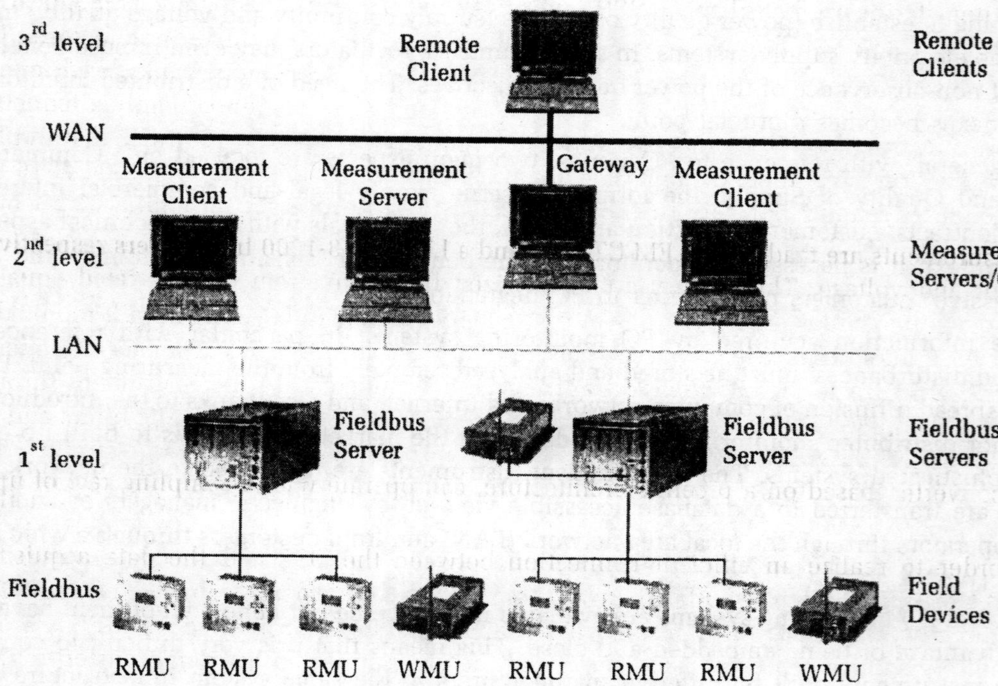

Fig. 1. *Architecture of the implemented distributed measurement system.*

In order to monitor points of the electrical system where a wire connection is impossible or too expensive, a wireless connection is used. The wireless connections assume that a wireless master modem (MM) is associated to the field-bus server and manages the entire data acquisition. Under the control of the field-bus server, the master modem sequentially interrogates, with addressed requests, all the wireless slave modems (SM) associated to the remote equipment and only one slave at the same time will provide its data. The single device can directly communicate with any other sensor node or, alternatively, the communication between two sensor nodes may involve a sequence of messages passing through the sensor chain by means of pair wise adjacent sensor nodes. As far as the connection to the master modem is concerned, it can be obtained both directly and, when the distance does not allow the connection, throwing a message passing technique. The reliable communication between the second and the third level is based on TCP/IP stack protocols.

5.2.1. Non-stationary Disturbance (NSD) Analyzer

The developed digital instrument implements a new method for a real time non-stationary disturbance analysis devoted to transients, dips, swells and interruptions. The method, fully described in, represents a valid alternative to the other methods for a fast and accurate detection, extraction and characterization of these disturbances. It is based on the idea of analyzing the acquired signals in the time domain sample by sample, utilizing both the concepts of sliding window and variable thresholds, according to the recent standards. The method is characterized by two fundamental properties : (*i*) the capability to take also into account stationary disturbances, like at harmonic and interharmonic components, (*ii*) the capability to limit the synchronization error with a suitable analysis and with a proper use of the sliding window.

The non-stationary disturbance (NSD) analyzer shown in Fig. 2, adopts a TMS320C6711 32-bit floating point DSP based on the very-long-instruction-word (VLIW) architecture. The special internal memory architecture, based on two independent memory banks, each one of 32 kB, allows two memory accesses within one instruction cycle and consequently a reduction of number of instructions. The processor operating frequency is equal to 150 MHz that implies a 6.6 ns instruction cycle time. On the same board, external memory banks for a total of 16 MB of synchronous dynamic RAM are directly connected to the processor through a 32 bit bus.

Measurements are made by a LEM CT 25-T and a LEM CV 3-1000 transducers respectively for current and voltage. The former is characterized by a maximum current rated equal to 25 A, an overall accuracy equal to 0.1% and a frequency bandwidth (-3dB) from 0 to 500kHz. The latter by a maximum voltage rated equal to 1000V, an accuracy of 0.2% and a frequency bandwidth (-1dB) up to 500kHz. These transducers are connected to the instrument through two-isolated channels. The two isolated signals are sent to low-pass anti-aliasing filters and are next passed to 12 bit A/D converter (THS1206) through two Sample-and-Hold circuits. The utilized converter, based on a pipeline architecture, can operate with a sampling rate of up to 6 MSps.

In order to realize an efficient connection between the DSP and the data acquisition system (DAS), a FIFO circular buffer that can store up to 16 conversion values is utilized. This buffer is directly interfaced to DSP through a 32-bit external memory interface (EMIF). A proper circuit controls both the A/D data acquisition and the FIFO buffer for the data transfer to EMIF interface. The performance of the NSD analyzer has been evaluated through experiments

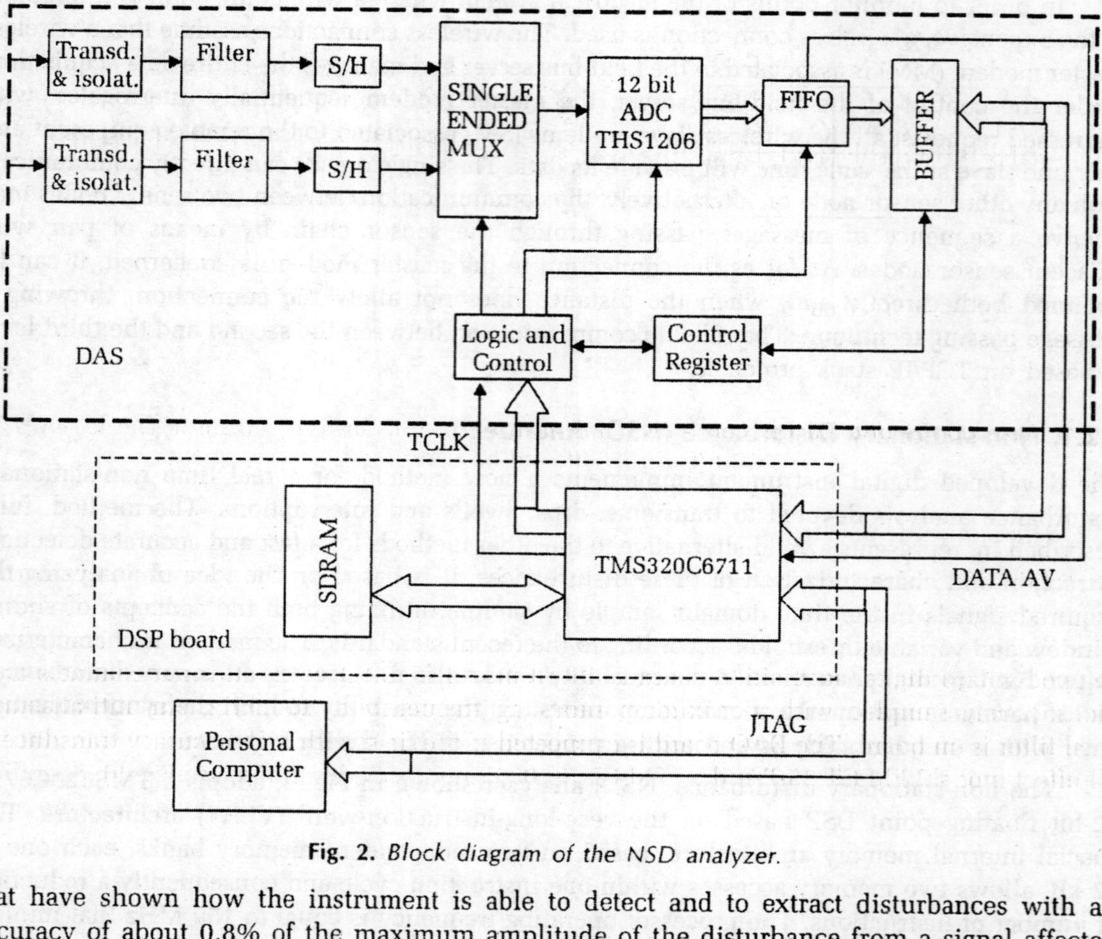

Fig. 2. *Block diagram of the NSD analyzer.*

that have shown how the instrument is able to detect and to extract disturbances with an accuracy of about 0.8% of the maximum amplitude of the disturbance from a signal affected by harmonic and interharmonic components according to the limits imposed by.

The instrumentation has a software interface, developed utilizing National Instruments CVI as a programming language, through which the operator can control it and read the results. The software interface runs on a personal computer with which the C6711 communicates via Joint Test Action Group (JTAG) interface. The data transfer is realized using real time data exchange (RTDX) library functions (see Fig. 3). This function allows transferring data between the digital instrument and a software application, running on the host, without interfering with the target application. The execution time is not influenced.

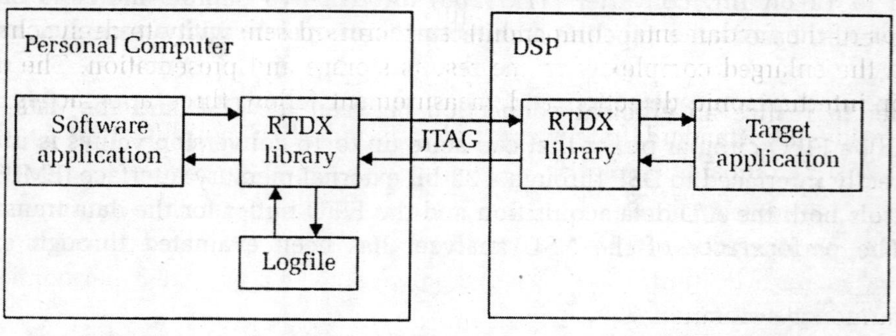

Fig. 3. *Use of RTDX library for target-software communication.*

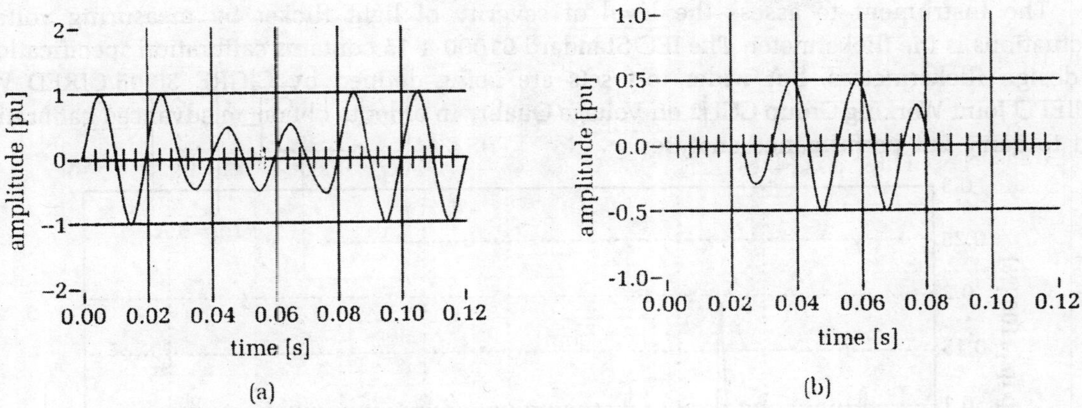

Fig. 4. *Detection of disturbances from a real time analysis by NSD analyzer, (a) signal affected by a sag (b) Signal affected by a swell.*

5.2.2. Harmonic, Interharmonic, Flicker, RMS and Peak Analyzer

A transportable measuring station (TRAMS) has been built up and tested in order to perform analysis to evaluate harmonic and interharmonic pollution for current and voltage, to assess flicker level and to monitor RMS and voltage peaks. It is based on a transportable computer equipped with a digital acquisition board at 16 bit model NI DAQ 4452; all input channels are synchronously sampled with a maximum sampling frequency of 200 kHz. An anti-aliasing digital filter is on board. The DAQ board is connected to the field with high accuracy transducer Hall-effect model LEM CT 25-T and a LEM CV 3-1000 previously described (see Fig. 5). Specific software for each analysis and monitoring has been developed in LabView environment accounting all the signal processing problems related to the considered phenomenon. For each parameter it is possible to perform a monitoring for very long periods (weeks, month).

Fig. 5. *Transportable measuring station (TRAMS).*

In particular, the analysis of the interharmonic distortion brings a lot of difficulties related to the change of waveform periodicity, to the very low amplitude of interharmonics of interest (few thousands of the fundamental component), an increased sensitivity to de-synchronization problems and the enlarged complexity in the results storage and presentation. The developed algorithms for interharmonic detection and measurement follow three approaches. The first approach is inside IEC standard framework and it is aimed to simplification and to obtain synthetic and easily comparable results.

The second approach performs a high accuracy harmonic synchronization in order to strongly reduce the interaction of these components on interharmonic evaluation. The third approach is based on a de-synchronized signal processing that performs harmonics filtering before interharmonic evaluation.

The instrument to assess the level of severity of light flicker by measuring voltage fluctuations is the flickermeter. The IEC Standard 61000-4-15 contains calibration specifications to design flickermeters, but wider test sets are being defined by CIGRE 36.05/CIRED WG 2/UIEPQ Joint Working Group CCU2 on Voltage Quality in order to obtain an advanced calibration and to carry out performance analyses.

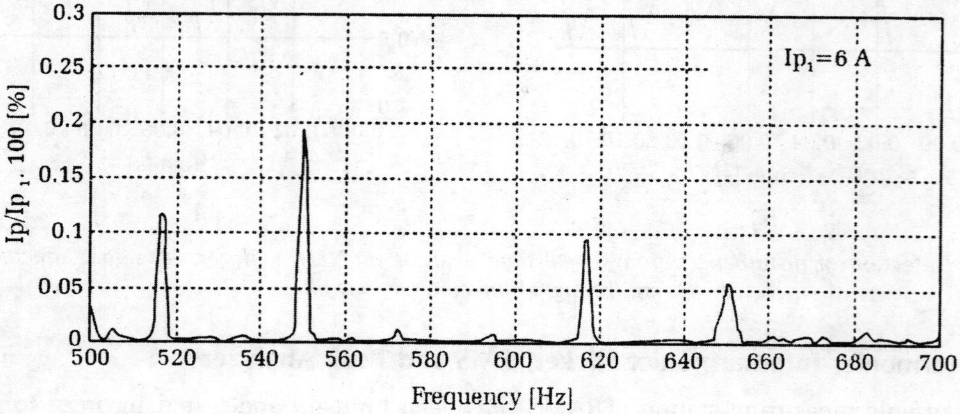

Fig. 6. *Example of high accuracy harmonic and interharmonic analysis on current.*

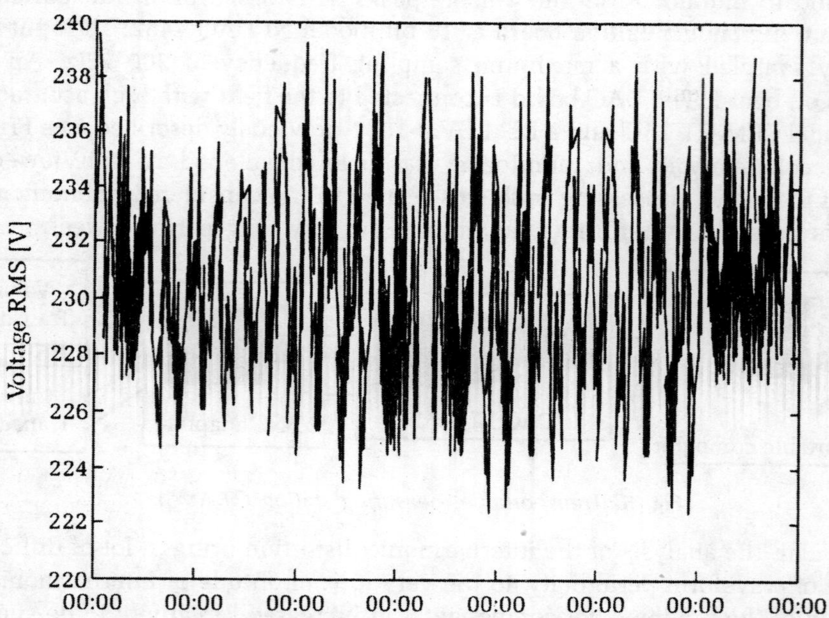

Fig. 7. *An example of measured RMS value of the voltage versus the time over one week.*

A digital implementation of standard flickermeter that is fully compliant with the largest test set has been implemented. Finally, a procedure for monitoring the peak value, the RMS and the fundamental component amplitude has been implemented. It measures the values of all the parameters each half cycle of the system fundamental component and, then, performs a statistical analysis of the obtained values.

5.2.3. PQ WEB Instrument

The hardware architecture of the proposed PQ web instrument, fully described in Fig. 8, has been designed in order to satisfy the following requirements : (*i*) low cost; (*ii*) continuous acquisition and data processing; (*iii*) wide remote communication capability. A block diagram of the sensor is shown in Fig. 8.

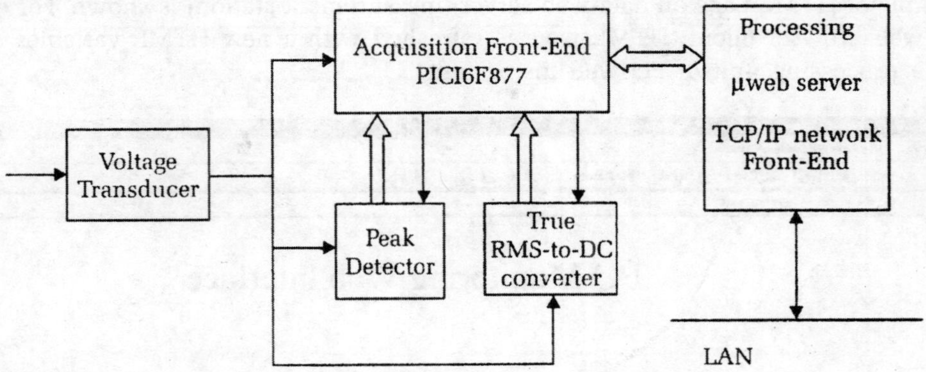

Fig. 8. *PQ web instrument architecture.*

The interface with the low voltage distribution system requires : (*i*) a galvanic insulation between the power line and the measuring circuit, in order to guarantee the safety of both the operator and measurement apparatus; (*ii*) a good linearity; (*iii*) a bandwidth suitable for the monitoring of fast transient events; (*iv*) small dimensions and weight. The voltage input transducer is based on an active attenuator. It allows (*i*) a nominal input voltage up to 600 V_{rms}; (*ii*) an overall accuracy of ± 0.3%; (*iii*) a thermal drift < 0.1%, (*iv*) a linearity < 0.2%; (*v*) a slew rate of 10 V/ms.

The device is able to measure some PQ parameters, in particular : voltage true rms amplitude, voltage peak amplitude, voltage total harmonic distortion, amplitudes of the fundamental frequency component and of the most important harmonics. The peak value can be used as a first index for the evaluation of transient disturbances. The rms value is helpful in the evaluation of different kinds of disturbances : (*i*) voltage short dips, with a width of less than one period of the fundamental frequency and a voltage shape distortion; (*ii*) voltage long dips, with a width of more than one period, without a voltage shape distortion; (*iii*) voltage flicker in compliance with the EN 61000-3-3 standard.

The input voltage acquisition should be performed without triggering, in order to avoid undesired acquisition gaps during fast transient events. Automatic data storage should be performed with a threshold-based technique, with a reference value set defined via software. To optimize the system performance, by increasing the measurement speed, the data acquisition, processing and communication tasks are assigned to different blocks : (*i*) voltage peak detector; (*ii*) true rms-to-dc converter; (*iii*) input signal acquisition and preprocessing; (*iv*) final data processing; (*v*) data transfer to the website managing unit. The apparatus has been provided with a communication link to a host PC, for the testing operations.

The data acquisition unit is based on a microcontroller, the Microchip PIC16F877. It performs the following operations : (*i*) continuous sampling, without triggering, of the outputs from the voltage transducer, the peak detector and the rms-to-dc converter; (*ii*) A/D conversion with a 10-bit vertical resolution; (*iii*) preprocessing of acquired raw data; (*iv*) data transfer to the website managing unit; (*v*) communication with the host PC during the apparatus testing.

The processing raw data and communication unit is based on the 80152 microcontroller. The measurement of the harmonic content of the acquired voltage has been implemented according to the EN 61000-4-7 standard, using the FFT-based approach. It carries out the final data processing and performs the results visualization by means of regeneration of the HTML pages in the web server. It also performs the control of the TCP/IP communication. In Fig. 9 an example of HTML page on data-web server (measurement station) is shown. For each access from a web browser client, HTML page is refreshed with a new HTML variables set coming from the processing unit of PQ web instrument.

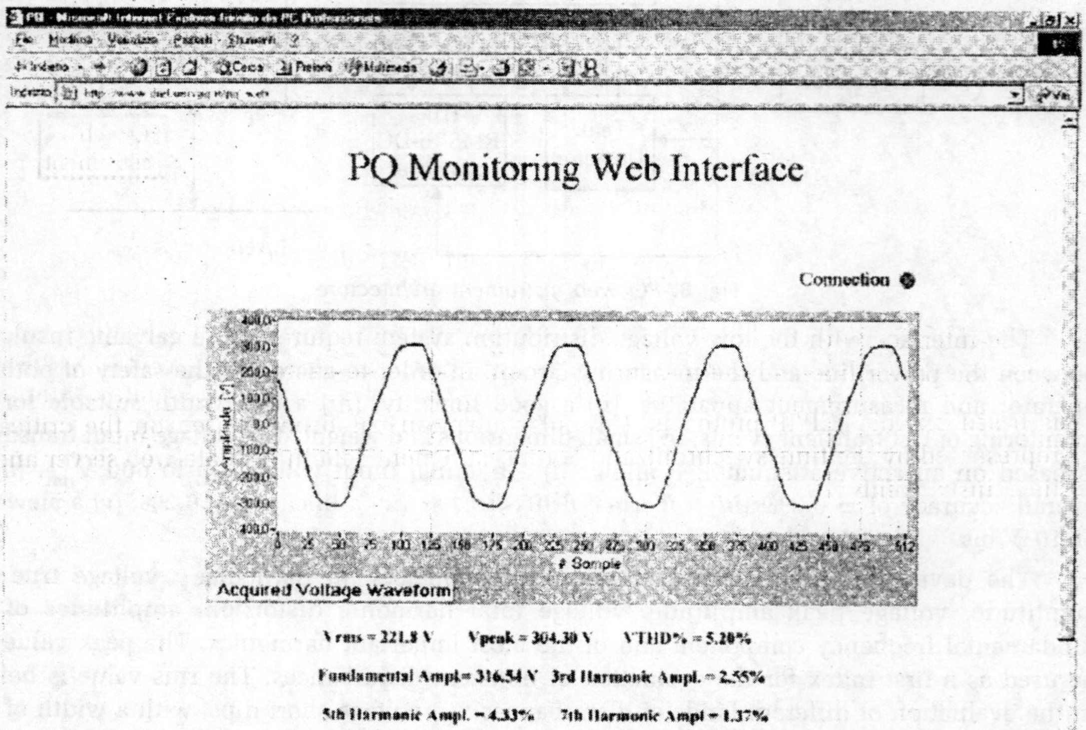

Fig. 9. *Example of generated HTML page by smart web sensor : steady state distorted voltage on low voltage sub-network.*

5.2.4. On-Field Monitoring

On-field monitoring were performed with reference to the distributed monitoring system reported in Fig. 10 that reflects the architecture proposed earlier. It is composed of the previously described advanced instruments and of commercial devices located at two distinct sites. Locally, the instruments are interconnected by a LAN and the acquired data are transferred to a local data-web server. The communication between the two LAN is possible by a WAN connection.

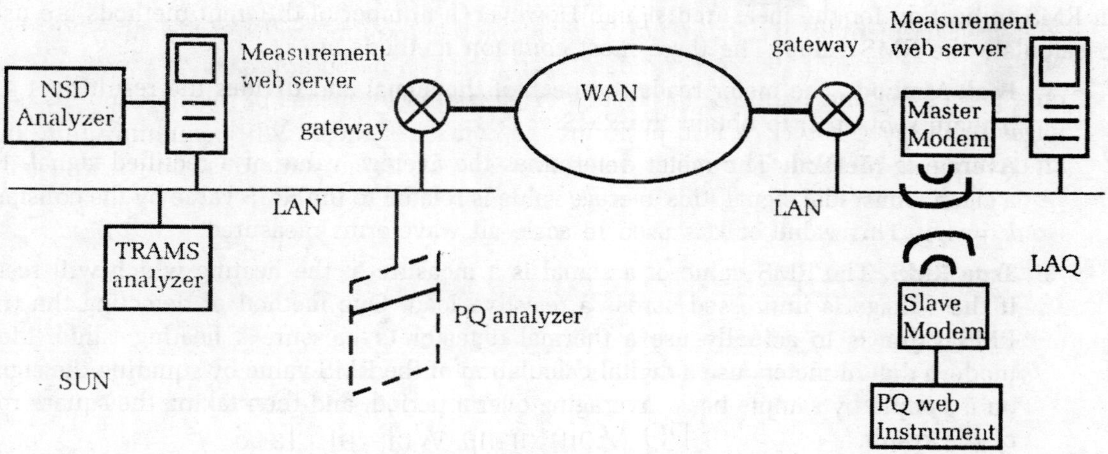

Fig. 10. *Distributed measurement system.*

The field monitoring devices considered are a commercial PQ, a NSD and a TRAMS analyzer located on SUN site and a PQ web instrument located on site. For each site, a personal computer operating as data-web server is utilized. On the data-web server the software to control specific devices and the software used for post-processing analyses (disturbance classification, statistical analysis, graphic functions etc.) run. The reliable data communication is based on the TCP/IP protocols. Particular attention has been focused on the critical point represented by the time synchronization of measurement data in the data-web server and in the field instruments.

5.3. TYPES OF EQUIPMENT FOR MONITORING POWER QUALITY

5.3.1. Multimeters or DMMs

After initial tests of wiring integrity, it may also be necessary to make quick checks of the voltage and/or current levels within a facility. Overloading of circuits, under and over-voltage problems, and unbalances between circuits can be detected in this manner. These measurements just require a simple multimeter. Signals to check include :

- Phase-to-ground voltages
- Phase-to-neutral voltages
- Neutral-to-ground voltages
- Phase-to-phase voltages (three phase system)
- Phase currents
- Neutral currents

There are numerous makes and models of Multimeters to choose from. A digital Multimeter with data storage is valuable because the waveform can be saved and analyzed. Multimeters in this category often have waveform analysis capability (energy calculation, spectrum analysis) also. In addition, the digital multimeters can usually be obtained with communications so that waveform data can be uploaded to a PC for additional analysis with a software package.

The most important factor to consider when selecting and using a multimeter is the method of calculation used in the meter. All of the commonly used meters are calibrated to give an RMS indication for the measured signal. However, a number of different methods are used to calculate the RMS value. The three most common methods are :

1. **Peak Method.** The meter reads the peak of the signal and divides the result by 1.414 (square root of 2) to obtain the RMS.
2. **Averaging Method.** The meter determines the average value of a rectified signal. For a clean sinusoidal signal, this average value is related to the RMS value by the constant, $k = 1.1$. This value of k is used to scale all waveforms measured.
3. **True RMS.** The RMS value of a signal is a measure of the heating which will result if the voltage is impressed across a resistive load. One method of detecting the true RMS value is to actually use a thermal detector to measure a heating value. More modern digital meters use a digital calculation of the RMS value by squaring the signal on a sample by sample basis, averaging over a period, and then taking the square root of the result.

Multimeters are used to measure voltages, currents, and resistances. Typical applications of multimeter measurements are line-voltage measurements and imbalance measurements. In making voltage and current measurements, generally we're interested in measuring the rms value of the waveforms. Some subtleties are associated with this measurement. In one type of multimeter, the input waveform is rectified and averaged, as shown in Fig. 11(*a*). This

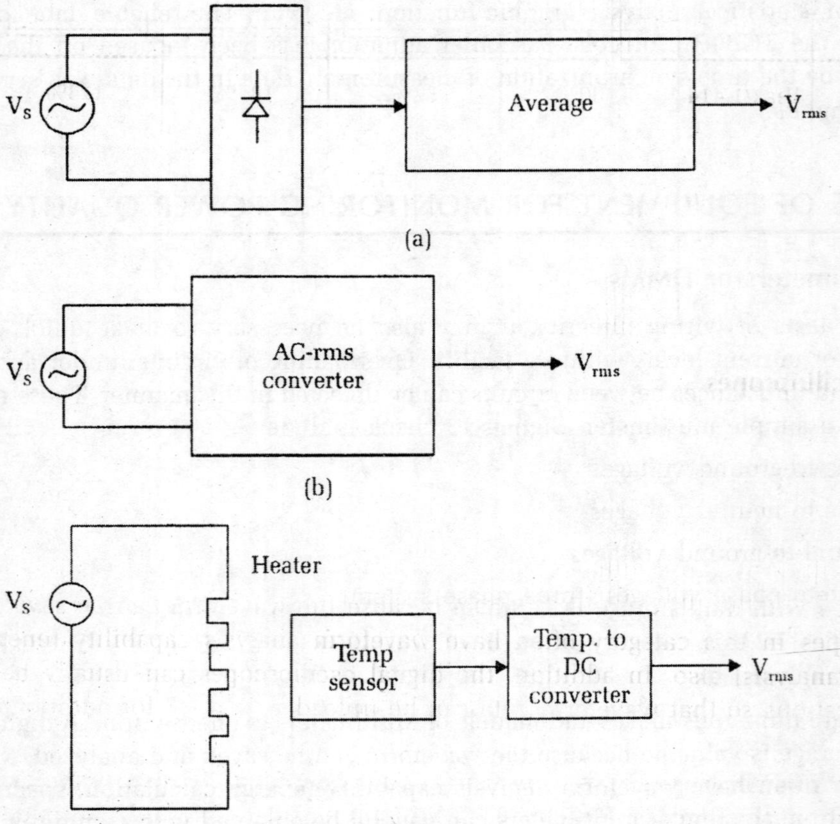

Fig. 11. *Multimeter types for making AC voltage measurements. (a) Rectify and average. (b) Analog computation. (c) Thermal.*

method works well for pure sine waves (where the rms value is the peak value divided by the square root of two), but for waveforms with harmonics, this method is prone to errors. Error is especially high for waveforms with a high crest factor. Several types of multimeters produce "true-rms" readings—that is, they accurately report the rms value of waveforms with harmonic distortion. One type of true-rms multimeter employs an rms-to-DC converter (Fig. 11(*b*)). Several IC manufacturers produce rms-to-DC converter integrated circuits.1 Another method of producing true-rms readings involves a thermal measurement method (Fig. 11(*c*)).

These different methods all give the same result for a clean, sinusoidal signal but can give significantly different answers for distorted signals. This is very important because significant distortion levels are quite common, especially for the phase and neutral currents within the facility. Table 1 can be used to better illustrate this point. Each waveform in Table 1 has an RMS value of 1.0 pu (100.0%). The corresponding measured value for each type of meter is displayed under the associated waveforms, per-unitized to the 1.0 pu RMS value.

Table 1 : Methods for Measuring Voltages and Currents with Multimeters

Meter Type	*Circuit*	*Sine Wave*	*Square Wave*	*Distorted Wave*	*Light Dimmer*	*Triangle Wave*
		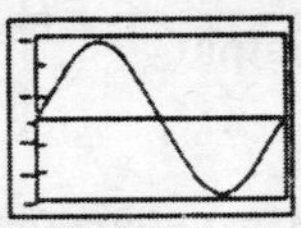	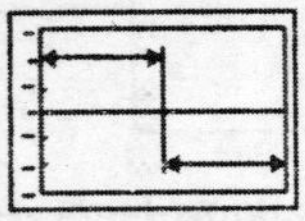	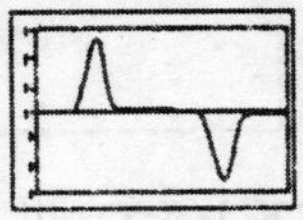	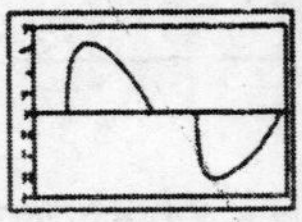	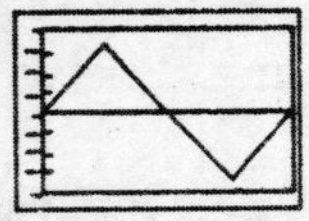
Peak Method	Peak/1.414	100%	82%	184%	113%	121%
Average Responding	Sine Avg. × 1.1	100%	110%	60%	84%	96%
True RMS	RMS Converter	100%	100%	100%	100%	100%

5.3.2. Oscilloscopes

An oscilloscope is valuable when performing real time tests. Looking at the voltage and current waveforms can tell a lot about what is going on, even without performing detailed harmonic analysis on the waveforms. You can get the magnitudes of the voltages and currents, look for obvious distortion, and detect any major variations in the signals.

There are numerous makes and models of oscilloscopes to choose from. A digital oscilloscope with data storage is valuable because the waveform can be saved and analyzed. Oscilloscopes in this category often have waveform analysis capability (energy calculation, spectrum analysis) also. In addition, the digital oscilloscopes can usually be obtained with communications so that waveform data can be uploaded to a PC for additional analysis with a software package.

Today's oscilloscopes have more than enough bandwidth to display the harmonics present in all power-quality events. For instance, the 20th harmonic of the line frequency in the U.S. is 1200 Hz, which is much lower than the bandwidth of even an inexpensive oscilloscope. Impulsive transients (like that caused by lightning strikes or power switching) typically last a few microseconds. It is well within the capability of oscilloscopes to measure and display these waveforms with good accuracy. Scopes are often used in conjunction with current probes2 to measure AC line currents. Often, differential voltage probe3 modules are used to measure the voltage difference between two points in a system. Many manufacturers make high voltage differential probes that are compatible with oscilloscope inputs. Typical specifications for differential probes4 are

- Bandwidth : 100 MHz
- Differential voltage : 5000 V
- Common-mode voltage : 2000 V

A typical system for making high voltage differential measurements is shown in Fig. 12.

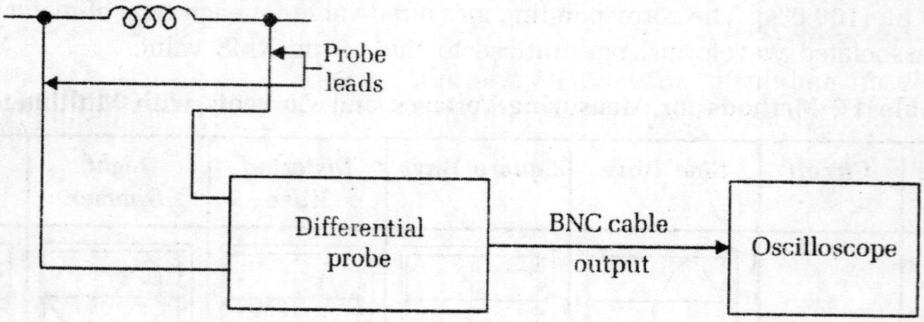

Fig. 12. *Measurement system for making high voltage differential measurements.*

5.3.3. Current Probes

Current probes come in several different types. Hall-effect current probes (like that used in the Tektronix P6042) detect all the way down to DC by utilizing the "Hall effect." Hall devices are semiconductors that generate an output voltage in response to a magnetic field created by a current. Hall probes have the advantage that they provide output at DC. Unfortunately, they are prone to drift and should be calibrated prior to use.

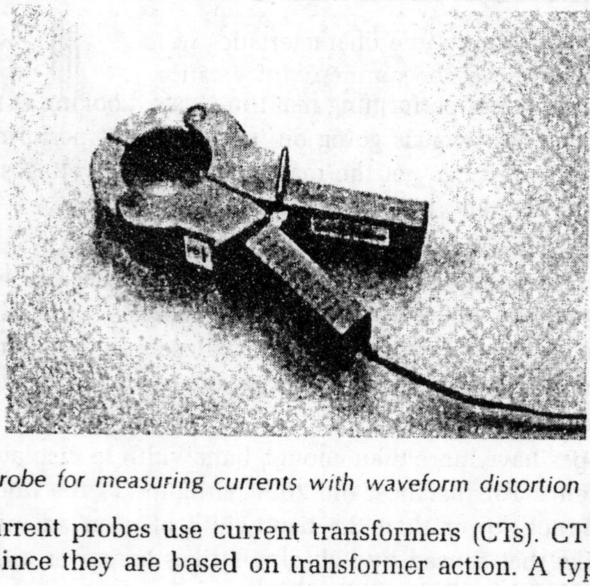

Fig. 13. *Current probe for measuring currents with waveform distortion due to harmonics.*

Other types of current probes use current transformers (CTs). CT current probes do not give an output at DC since they are based on transformer action. A typical test setup using a

CT and an oscilloscope is shown in Fig. 14. CT current probes give a conversion ratio (volts per amp) when terminated in the proper resistance—in most cases 50 ohm.

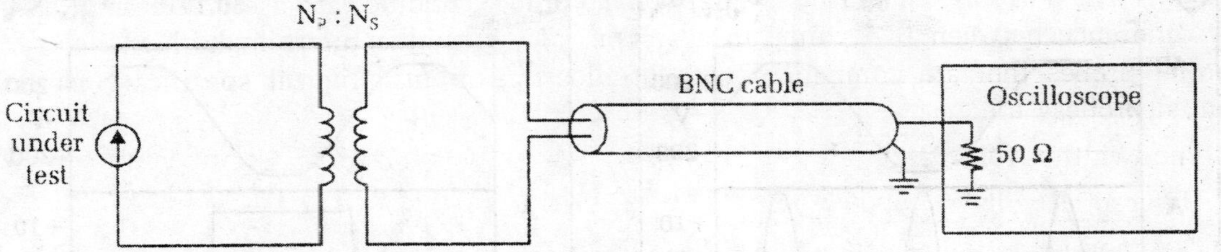

Fig. 14. *A measurement system for making high current measurements using a CT and an oscilloscope.*

5.3.4. Disturbance Analyzers

Disturbance analyzers and disturbance monitors form a category of instruments which have been developed specifically for power quality measurements. They typically can measure a wide variety of system disturbances from very short duration transient voltages to long duration outages or under-voltages. Thresholds can be set and the instruments left unattended to record disturbances over a period of time. The information is most commonly recorded on a paper tape but many devices have attachments so that it can be recorded on disk as well. There are basically two categories of these devices :

1. **Conventional analyzers** that summarize events with specific information such as over/under voltage magnitudes, sags/surge magnitude and duration, transient magnitude and duration, etc.
2. **Graphics-based analyzers** that save and print the actual waveform along with the descriptive information which would be generated by one of the conventional analyzers. It is often difficult to determine the characteristics of a disturbance or a transient from the summary information available from conventional disturbance analyzers. For instance, an oscillatory transient cannot be effectively described by a peak and a duration. Therefore, it is almost imperative to have the waveform capture capability in a disturbance analyzer for detailed analysis of a power quality problem. However, a simple conventional disturbance monitor can be valuable for initial checks at a problem location.

The most recent instruments combine limited harmonic sampling and energy monitoring functions with complete disturbance monitoring functions as well . The output is graphically based and the data is remotely gathered over phone lines into a central database. Statistical analysis can then be performed on the data. The data is also available for input and manipulation into other programs such as spreadsheets.

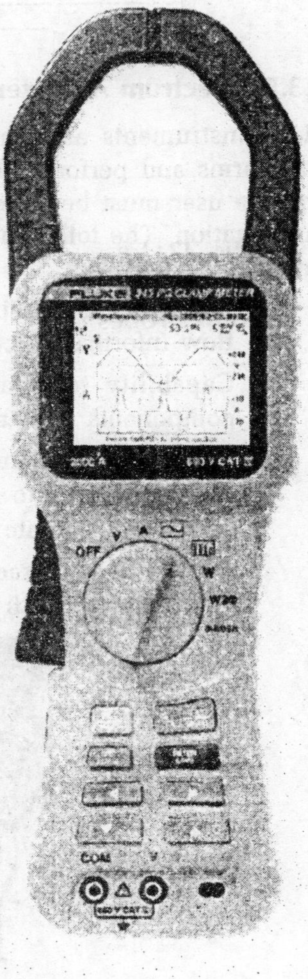

Fig. 15. *Handheld disturbance analyzer.*

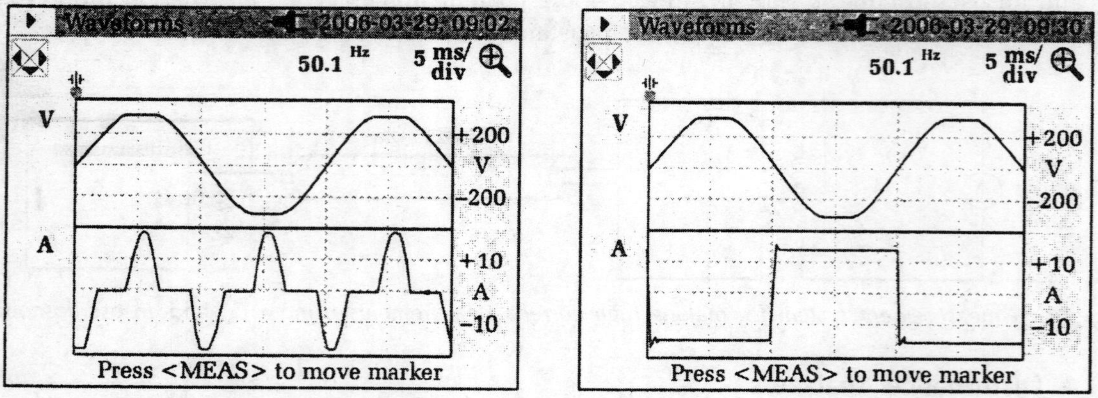

Fig. 16. *The dual waveform display on a disturbance analyzer shows both voltage and current waveforms.*

5.3.5. Spectrum Analyzers and Harmonic Analyzers

Many instruments and online monitoring equipment now include the capability to sample waveforms and perform FFT calculations. The capabilities of these instruments vary widely and the user must be careful that the accuracy and information obtained is adequate for the investigation. The following are some basic requirements for harmonic measurements used to investigate a problem :

- Capability to measure both voltage and current simultaneously so that harmonic power flow information can be obtained.
- Capability to measure both magnitude and phase angle of individual harmonic components (also needed for power flow calculations).
- Synchronization and a high enough sampling rate for accurate measurement of harmonic components up to at least the 37th harmonic (this requirement is a combination of a high sampling rate and a sampling interval based on the 50 Hz fundamental).
- Capability to characterize the statistical nature of harmonic distortion levels (harmonics levels change with changing load conditions and changing system conditions).

Fig. 17. *Handheld harmonic analyzer showing voltage leads and current probe for voltage and current harmonic measurements.*

Harmonic distortion is a continuous phenomena. It can be characterized at a point in time by the frequency spectrums of the voltages and currents. However, for proper representation, measurements over a period of time must be made and the statistical characteristics of the harmonic components and the total distortion determined.

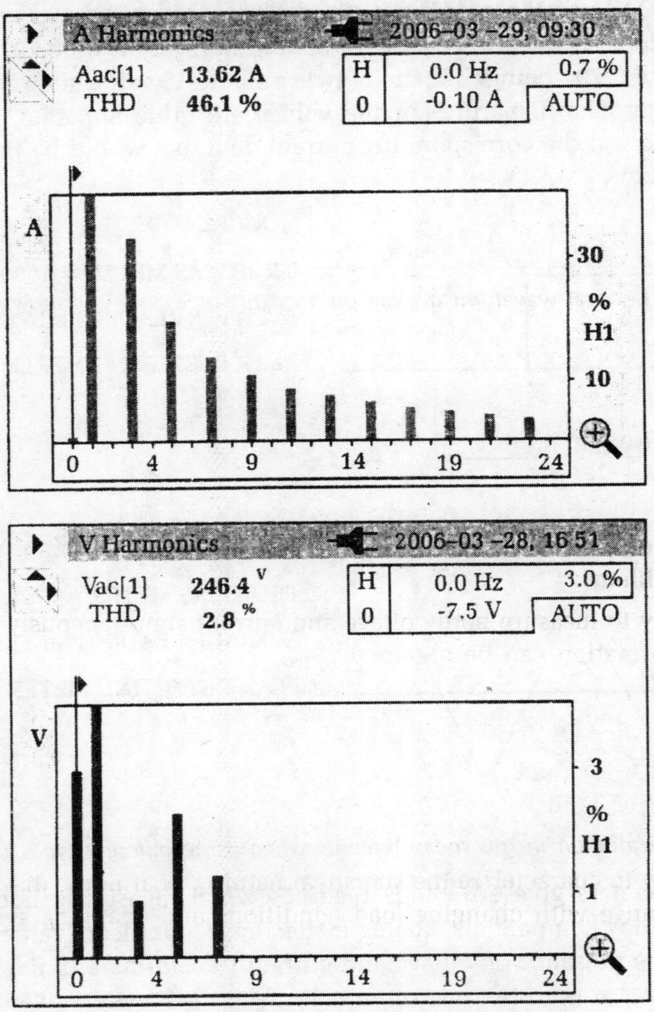

Fig. 18. *Graph harmonics for voltage and current.*

5.3.6. TRUE RMS METERS

The term true RMS is commonly used in power quality applications. What are true RMS meters ? As we saw in previously the RMS value of the current or voltage can be substantially different from the fundamental component of the voltage or current. Using a meter that measures average or peak value of a quantity can produce erroneous results if we need the RMS value of the waveform. For waveforms rich in harmonics, the average and peak values would be considerably different than waveforms that are purely sinusoidal or close to sinusoidal. Measuring the average or peak value of a signal and scaling the values to derive a RMS value would lead to errors.

For example, consider a square wave of current as shown in Fig. 19. The average and peak reading meters indicate values of 111 A and 70.7 A RMS current, respectively. The square waveform has an average value of 100 A. The peak value of the waveform also has a value of 100 A. In order to arrive at the RMS value, the 100 A average value is multiplied by 1.11. The ratio between the RMS and the average value of a pure sinusoidal waveform is 1.11. The peak reading meter would read the 100 A peak value and multiply it by 0.707 to arrive at the RMS value of 70.7 A, with 0.707 being the ratio between the RMS value and the peak value of a pure sinusoidal waveform. The disparities in the values are quite apparent. Fig. 19 also shows a triangular waveform and the corresponding current data that would be reported by each of the measuring instruments.

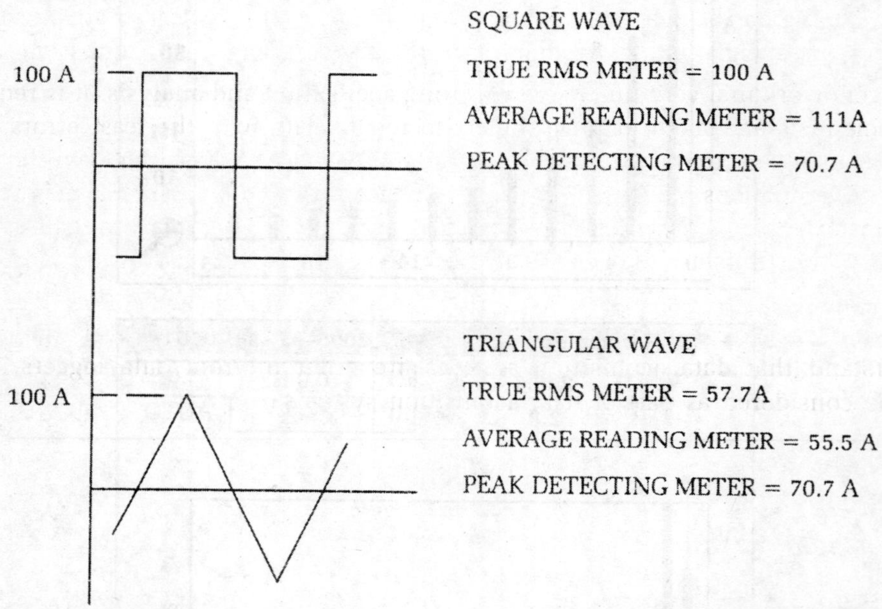

Fig. 19. *Variation in rms measurements when using different types of meters.*

Analog panel meters give erroneous readings when measuring nonsinusoidal currents. Due to higher frequency components, analog meters tend to indicate values that are lower than the actual values. The presence of voltage and current transformers in the metering circuit also introduces additional errors in the measurements. True RMS meters overcome these problems by deriving the heating effect of the waveform to produce an accurate RMS value indication. After all, RMS value represents the heating effect of a voltage or current signal. Most true RMS meters do not provide any indication of the waveform of the quantity being measured. To accomplish this, the meters require high-frequency signal sampling capability. The sampling rate should satisfy Nyquist criteria in order to produce reasonably accurate results. Some benchtop RMS meters do have the sampling capability and ports to send the information to a computer for waveform display.

5.3.7. Data/Chart Recorders

A data/chart recorder is an electronic instrument that keeps track of various measurements required in industrial and laboratory environments. Broadly speaking, it is used to record :

- Process measurements for such variables as temperature, pressure, flow, pH, and humidity

- Scientific and engineering data for applications such as testing and diagnostics, statistical analysis, and other laboratory work that requires a graphic or digital record of variables

There are scores of data recorder models on the market, ranging from Rs. 4200 standalone chart recorders that use paper and a pen to measure a single input, such as temperature, to Rs. 200000-plus high-speed models that accept 30 inputs. These, however, are the extremes—gone are the days when most data recorders could be classified as simple, midrange, or complex. In recent years, for example, conventional analog chart recorders have hybridized with data loggers, and PC-based functions have been added to most of the product categories outlined below.

Data loggers or transducers are used to record real life data of life science and research projects. Data logging has become ever so important that numerous research organizations and medical institutions spend millions of dollars in **data loggers** and data acquisition systems.

Data loggers work in a very unique way. During acquisition and analysis, it is required for researchers, scientists and medical practitioners to record data with the least errors possible. This is done by using electrodes and transducers that are capable of reading even the minutest of movements and variations. This helps in effectively recording motion, which is then transferred into digital or electronic signals. These electronic signals are stored into acquisition systems, which can then be transferred or downloaded into a computer for future analysis and interpretation.

The use of data loggers and data acquisition systems have increased over the years. One should understand that data acquisition systems are different from data loggers, but data loggers can be considered as part of data acquisition systems.

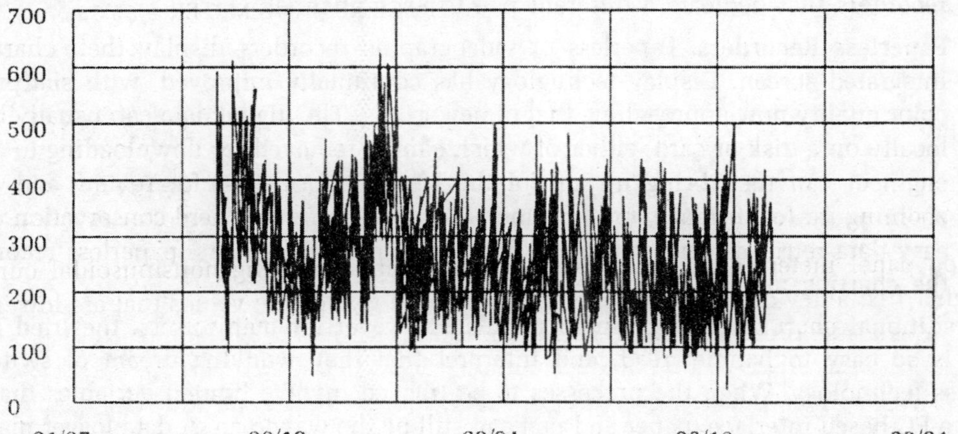

Fig. 20. *Current data from a data logger for one month of tests.*

While choosing data acquisition systems, you should take extreme care in investing in the best data logging systems available in the market. Cost shouldn't be a real consideration for data recording, since it involves numerous working hours. However, if cost indeed is a consideration, don't always go in for cheap systems, you can get pretty good systems with average pricing.

5.3.7.1. Types of Data/Chart Recorders

- **Strip Chart Recorders.** Strip chart recorders consist of a roll or strip of paper that passes linearly beneath one or more pens. As the signal changes each pen's deflection records the process being measured in the form of a chart. Well suited to recording of continuous processes, strip chart recorders are commonly used in both laboratory and process-measurement applications. For future reference, sections of the paper can be torn off and archived.

- **Circular Chart Recorders.** A circular chart recorder records data on a paper disc rotated beneath one or more pens, which, as in a strip chart recorder, deflect with fluctuating electrical signals. The difference is that the resulting chart is circular rather than linear. Circular chart recorders are ideal for batch processes that operate within a known timeframe. They can be configured so that each rotation of the chart covers a standard time period—1 hour, 24 hours, 7 days, etc. Some recorders will also accommodate non-standard periods. The advantage of a circular chart is that, at a glance, it gives a complete history of one or several variables over the specified period.
- **XY Recorders.** These recorders accept two inputs and create a chart or graph that displays the activity of one set of data against another. They are useful for determining relationships between the two inputs; for example, an XY recorder might be used in a chemical process to monitor the effect of temperature on pressure.
- **Hybrid Recorders.** A hybrid recorder can function not only as a recorder but also as a data logger, a device that accepts an input, or channel, that is fed into an analog-to-digital converter. The results are stored digitally or printed out as a series of time-stamped values. A hybrid recorder can combine analog trend representations and digital information on the same chart paper. Hybrid recorders typically come in multichannel designs, with one print head normally handling all channels. They offer a cost-effective solution for multichannel processes, though they have a slower response time than recorders that dedicate a different pen to each channel.
- **Paperless Recorders.** Paperless or videographic recorders display their charts on an integrated screen. Display technology has continually improved, with sharpness and color quality now approaching that of newer PCs. The digital data can usually be stored locally on a disk or card, either of which can be removed for downloading to a PC. An engineer can then bring up any of the stored information for review and analysis, zooming in, for instance, on the time of a process upset. Where conservation of paper, easy data retrieval, and sharing of data are important benefits, paperless recorders are the charting method of choice.

Traditional chart recorders remain popular, however for many users, the tried and true output is so easy to handle, read, and interpret that they wouldn't dream of switching to paperless technology. When the processes to be tracked involve limited variables that do not require a PC-based interface, paper and pen can still be the way to go. A data logger may accept a greater number of inputs, but only a recorder provides a truly continuous trend display of a variable's change with time. With advanced electronics such as microprocessors, improved ergonomics, and new user-interface features, the latest models are anything but low tech.

5.4. ANALYZING POWER QUALITY MEASUREMENT DATA

Analyzing power quality measurements has become increasingly more sophisticated within the past few years. It is not enough to simply look at RMS quantities of the voltage and current. Disturbances that occur on the power system have durations in the milli-second time frame, equipment is more sensitive to these disturbances, and there is more equipment connected to the power systems that cause disturbances or power quality problems. For these reasons, it is often necessary to continuously monitor system performance and characterize possible impacts of disturbances.

The data analysis system must be flexible enough to handle data from a variety of monitoring equipment and maintain a database that can be used by many different applications. The concept is illustrated in Fig. 21. Different types of power quality variations require different types of analysis to characterize system performance. Some examples are given in the following sections. With a flexible system, these applications can be customized to individual user needs.

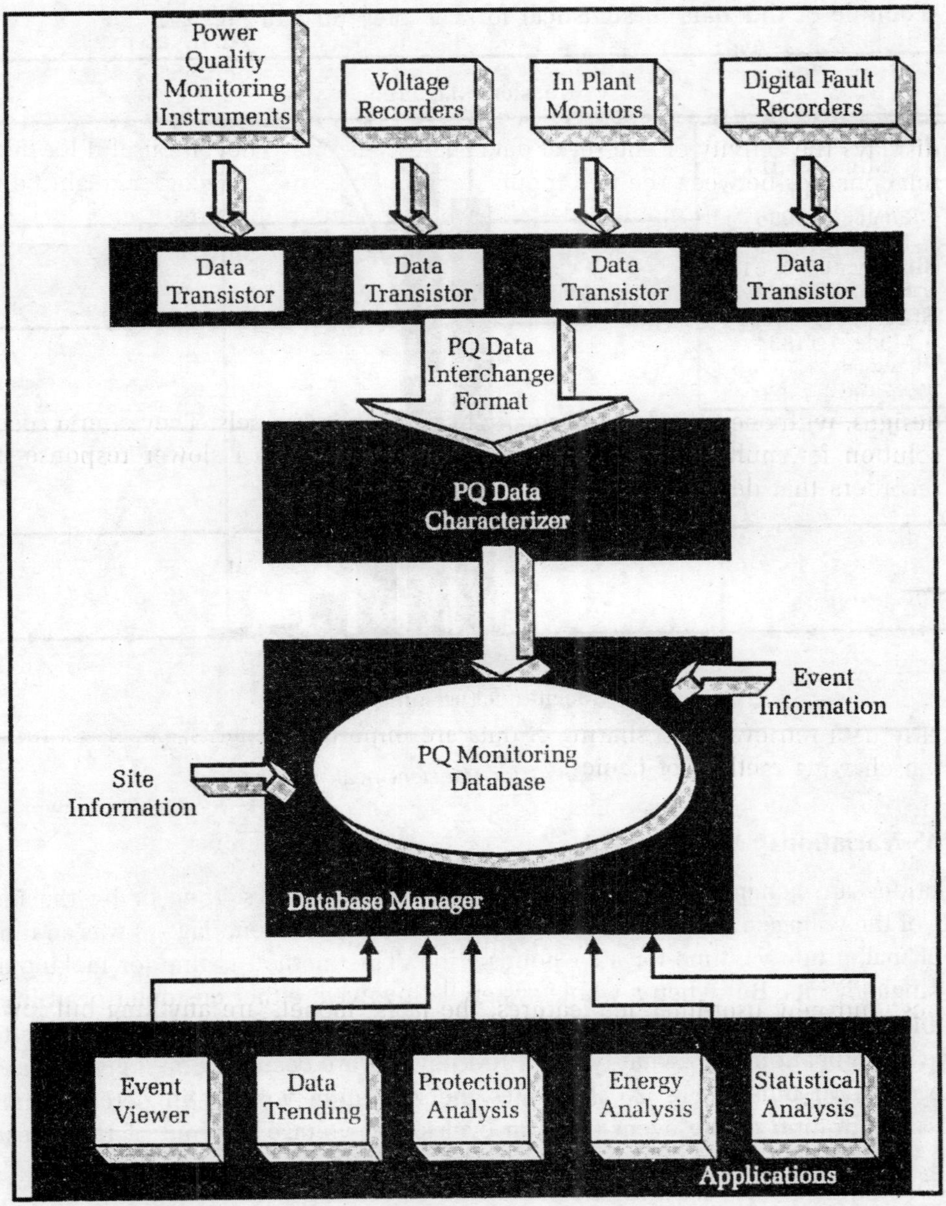

Fig. 21. *Example data analysis system.*

5.4.1. Transients

Transients are normally characterized by the actual waveform, although summary descriptors can also be developed for :

- Peak magnitude
- Primary frequency
- Time of occurrence
- Rate of rise

An example of this data in statistical form is presented in Fig. 22.

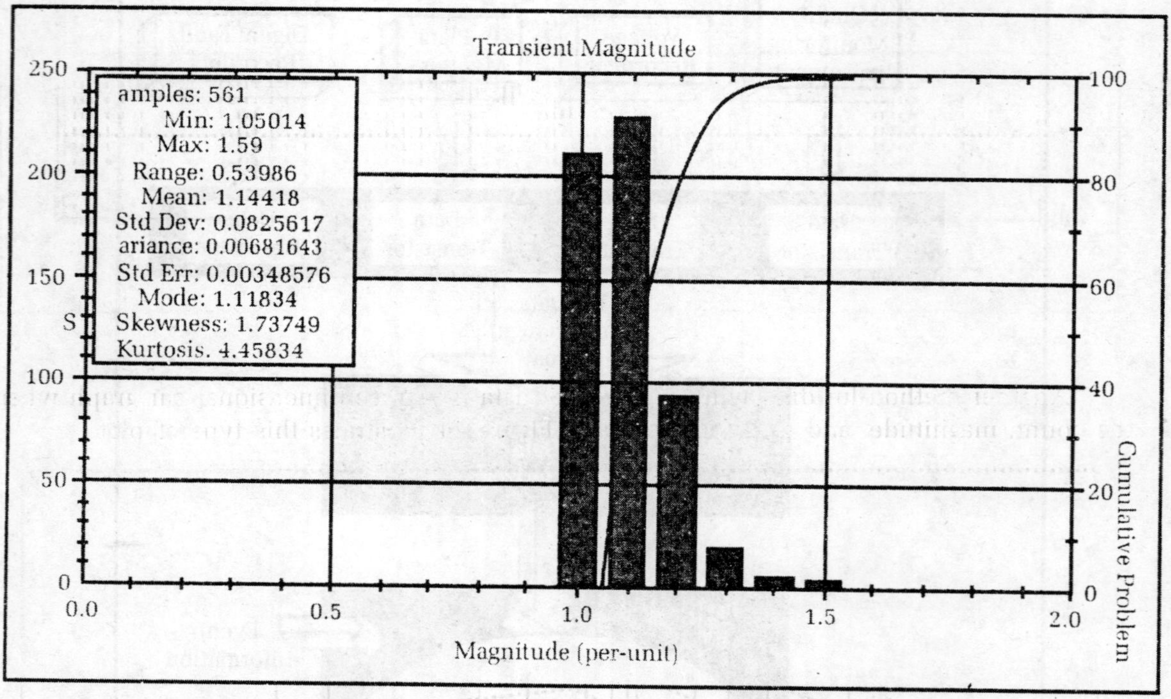

Fig. 22. *Bar chart for transient peak voltage.*

5.4.2. RMS Variations

RMS variations are generally characterized by the RMS value vs. time or by the minimum magnitude of the voltage during the event vs. the duration of the event. Fig. 23 was an illustration of a plot of magnitude vs. time for a 24 hour period. This method is fine for looking at single sites and single events. But when a whole system is involved, either customer or utility, it may be preferable to look at a range of events (e.g., one month, one year, etc.) for multiple sites. This would give an indication as to what types of RMS events are occurring on a given system. The magnitude duration plot in Fig. 23 illustrates the minimum voltage (in percent) during the event and the duration of the event (time in cycles that voltage was out of the thresholds).

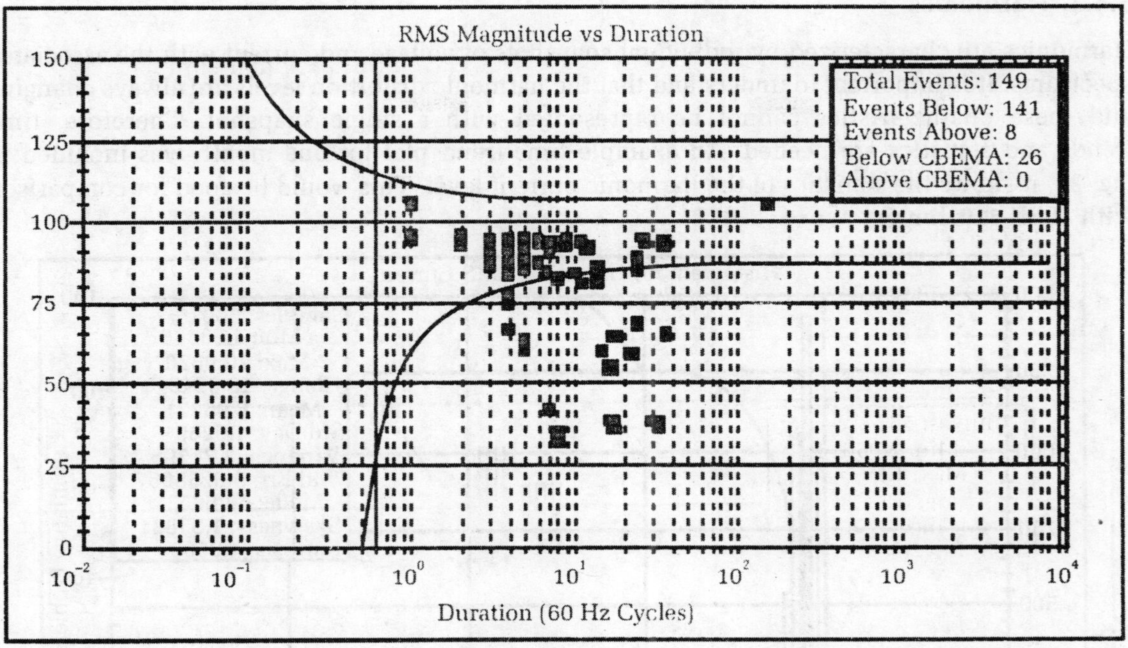

Fig. 23. *Example magnitude duration plot.*

Another method for displaying this type of data is a three dimensional bar graph where the count, magnitude, and duration is shown. Figure 24 illustrates this type of plot.

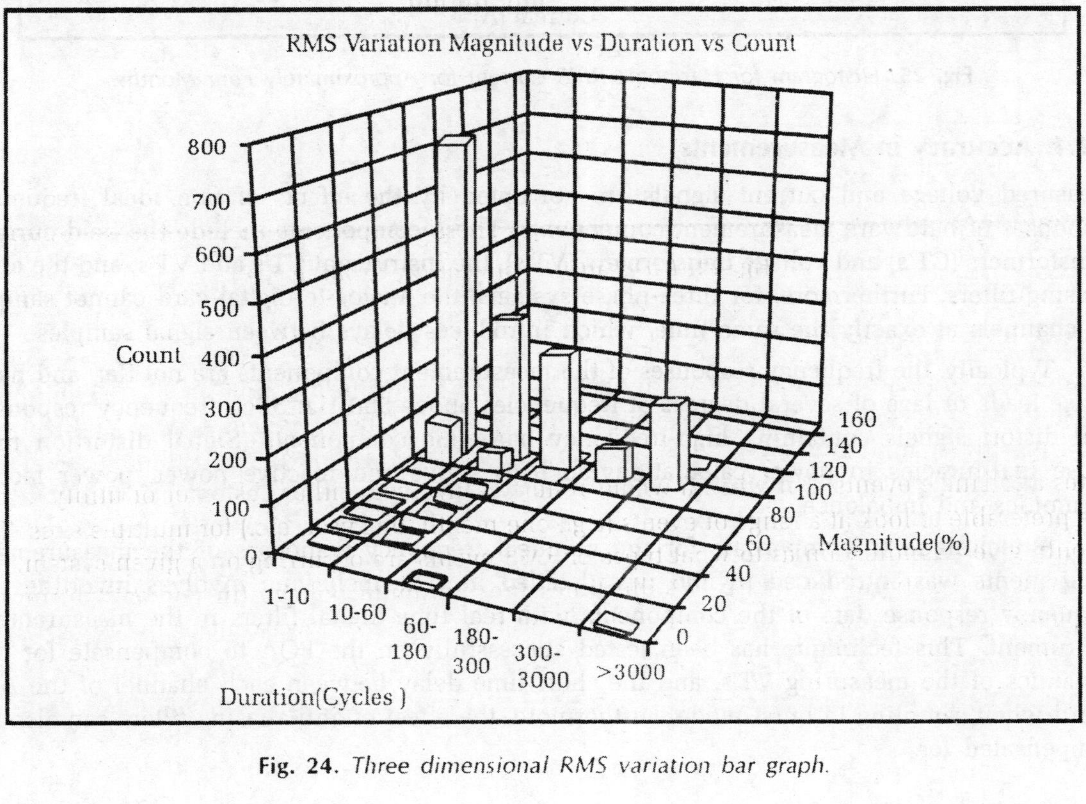

Fig. 24. *Three dimensional RMS variation bar graph.*

5.4.3. Harmonics

Harmonics are characterized by individual snapshots of voltage and current with the associated spectrums. It is important to understand that the harmonic distortion levels are always changing and these characteristics cannot be represented with a single snapshot. Therefore, time trends and statistics are needed. An example time trend plot for one month was included in Fig. 25, it shows the statistics of the harmonic current level. This would be good for comparison with IEEE-519 limits.

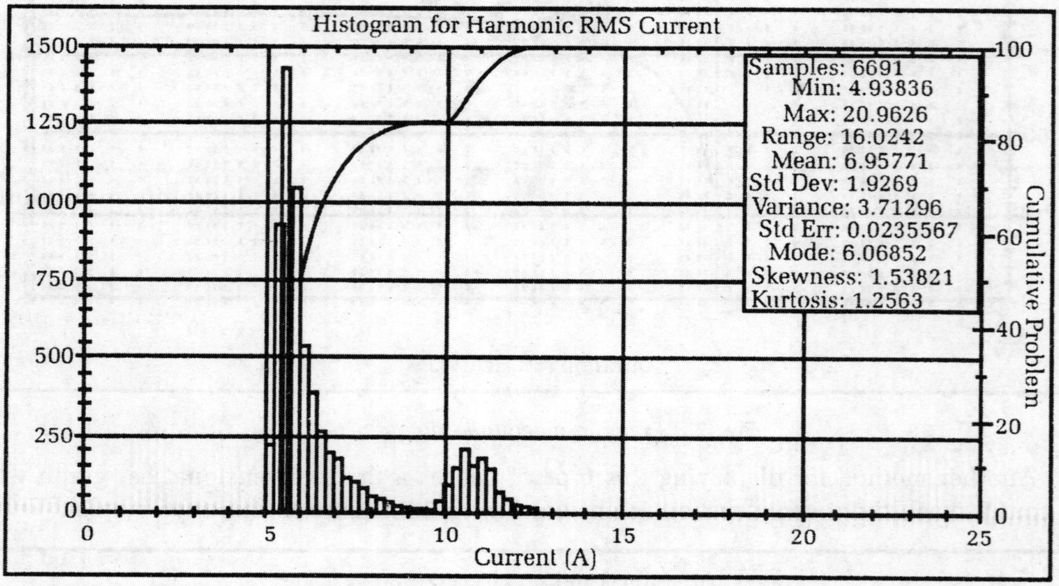

Fig. 25. *Histogram for Harmonic RMS Current for Approximately Four Months.*

5.4.4. Accuracy in Measurements

Measured voltage and current signals are corrupted by the effects of non ideal frequency responses of hardware measurement components. These components include the field current transformers (CT's) and voltage transformers (VT's), the instrument CT's and VT's, and the anti-aliasing filters. Furthermore, for three-phase systems, the analog-to-digital card cannot sample all channels at exactly the same time, which introduces delays between signal samples.

Typically, the frequency responses of the measurement components are not flat, and have phase leads or lags of several degrees at frequencies above 500 Hz. Such frequency responses will distort signals containing high-frequency harmonic components. Signal distortion may cause inaccuracies in power calculations such as active and reactive power, power factor, harmonics and imbalance.

A technique to compensate for the nonideal frequency responses of the measurement components was introduced in and was detailed in. The technique involves inverting the frequency response data of the components with real time digital filters in the measurement instrument. This technique has been tested successfully on the PQA to compensate for the dynamics of the measuring VT's, and the short time delay between each channel of the A/D board when sampling in burst mode. Furthermore, the effect of anti-aliasing filters can also be compensated for.

Compensating filters that approximately invert the frequency responses of the measurement components are calculated using frequency-domain interpolation techniques. These techniques are used to interpolate the inverse of the complex frquency-response data with a proper, stable transfer function on the imaginary axis. This transfer function can be implemented as a digital filter in the PQA, so that approximate cancellation takes place and the resulting combined frequency response is close to unity magnitude and zero phase in the frequency band of interest. The interpolating transfer function is mapped into the z-domain via the bilinear transformation with the appropriate sampling period. The resulting discrete-time filter is implemented as a first-order, discrete-time state-space matrix equation in the PQA's software.

5.5. PQA FEATURES

The PQA consists of an Intel-based processing unit connected to its own instrument current transformers and voltage transformers, via a 16-bit analog-to-digital card. Software modules have been developed to perform real-time calculations in a multitasking environment. The multitasking environment combines power quality measurements that, up to now, required many meters. In addition, the user may configure calculation output averaging intervals to match individual requirements or specific contract clauses—a feature that most other meters do not provide. Use of an Intel-based processing unit rather than custom hardware provides flexibility and an economical way to monitor all power quality needs.

The PQA was designed in a modular fashion, allowing calculations to be easily added or removed to suit individual requirements. New standards and custom calculations can be implemented easily with such a modular approach. The power quality calculations that have been developed for the PQA, to date, have been grouped together into three separate modules; the basic electrical module (BEM), the advanced electrical module (AEM), and the arc furnace module (AFM).

Input signals and calculation results may be accessed or viewed while calculations are being performed. Data access is achieved by reading output data files to a local machine, or by connecting to a remote machine through a TCP/IP network connection. With the use of TCP/IP, worldwide access to the output of any PQA can be obtained via the Internet. More private and secure access can be established on an internal network. As the PQA was developed to be useful for monitoring remote sites with only modem or low bandwidth WAN/Internet connections, care was taken to keep network traffic to a minimum. Data is compressed, and transferred data is cached locally. Thresholds and event detection may be used to have data transfer focused on events.

Calculations can be pre-configured to write to disk, either continuously, or upon the detection of pre-selected events, thereby conserving disk space. In the latter case, the meter may be left unattended for weeks or months, and later provide a record of events that have occurred during that time. The output from the PQA calculations can be averaged over overlapping, or non-overlapping demand intervals (sliding or fixed demand intervals respectively). Both the demand interval and overlap interval (see Fig. 26) may be set to any number of electrical cycles. The size of the demand and overlap intervals can be as small as 1 cycle, or as large as required; hours or even days. In the example in Fig. 26, the demand and overlap intervals are set to 3 and 1 cycle respectively.

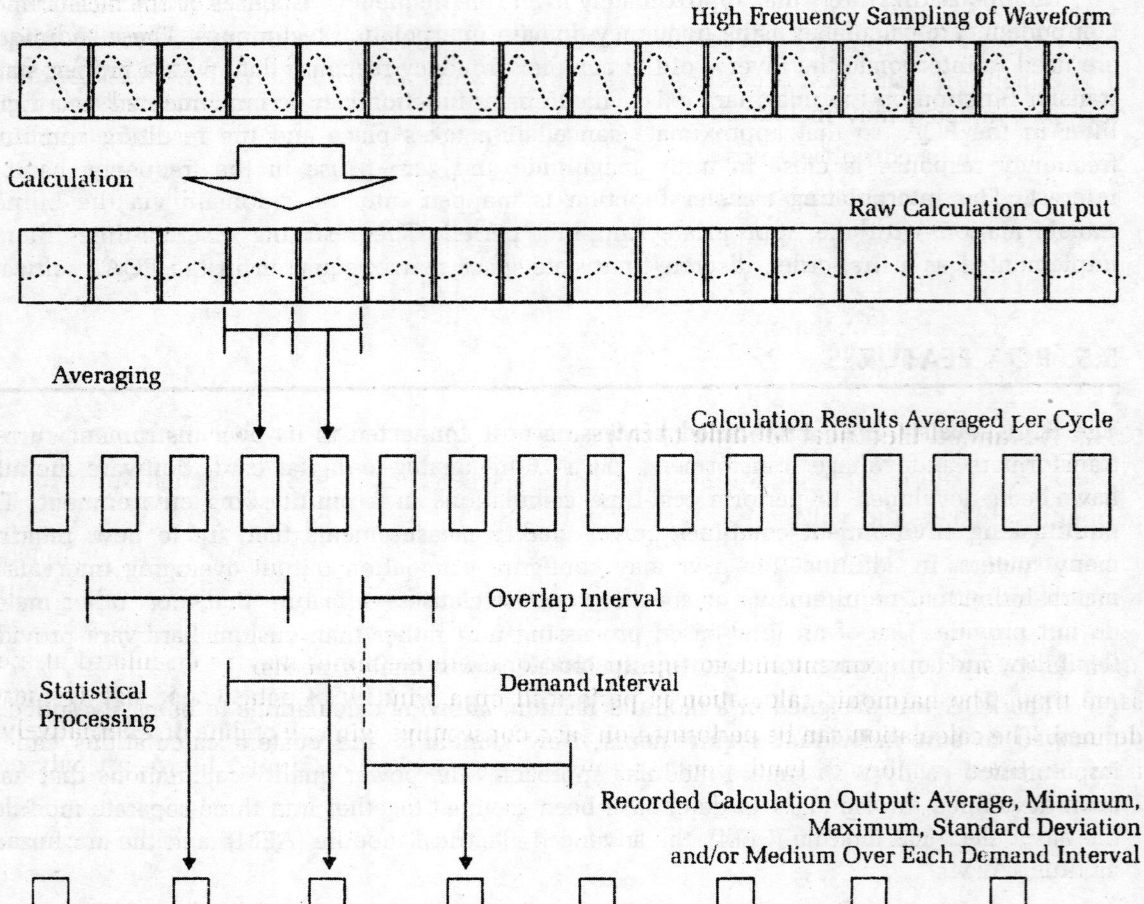

Fig. 26. *Output averaging of calculation data.*

For each demand interval, the user can choose to have any of the average, minimum, maximum, median or standard deviation calculated. Calculations can be configured to output for every demand interval, or only for demand intervals in which the average value of the output is outside of a user-selected range. The averaging and statistical evaluation described above provides unusual flexibility in data trending. As an alternative to storing large data trends, a facility to store the statistical cumulative probability function is provided. The cumulative probability function can be computed on a daily and/or weekly basis, and have only a few percentiles retained, in order to compare the disturbance levels for the day or week to standard compatibility levels.

PQA calculations provide raw output at the same rate as waveform sampling. Averaging of the data is done to reduce the output data to one sample per cycle. Using a user-specified demand and overlap interval, further statistical processing is done on each demand interval to produce the recorded calculation output.

5.5.1. Basic Electrical Module (BEM)

The Basic Electrical Module provides standard power calculations. With flexible user-defined output capabilities, the BEM is useful for evaluating a wide range of electrical problems online. Problems evaluated may include :

1. Fault-induced voltage dips
2. Short-term RMS variations
3. Long-term RMS variations
4. Phase imbalance
5. Asymmetrical operation
6. Low power factor

5.5.2. Advanced Electrical Module (AEM)

The Advanced Electrical Module provides detailed waveform calculations. Calculations include harmonics, flicker, and voltage dip classification.

Harmonics : The harmonic calculation provides real time monitoring of the complete frequency spectrum as well as the individual harmonic components. The input signal to this calculation may be a voltage or a current. Since many harmonic calculations may be executed simultaneously, both current and voltage spectra for all three phases may be calculated at the same time. The harmonic calculation is performed on a window of data whose size is user-defined. The calculation can be performed on each consecutive window of data or, alternatively, a user-defined number of fundamental frequency cycles can be skipped between analysed windows.

To prevent spectral leakage, the input data may be windowed before the harmonic calculation is performed. Standard window functions implemented include Kaiser-Bessel, Hamming, Hanning and Exponential. A comprehensive discussion of data windowing can be found in. With each analysed window of data, the following parameters are calculated :

1. Total Harmonic Distortion (THD) using the fundamental component,
2. Total Demand Distortion (TDD) using a user-specified nominal value,
3. Telephone Interference Factor (TIF),
4. Crest Factor,
5. Form Factor,
6. Peak Volts/Amps of the fundamental component.

The harmonic calculation outputing facility is very flexible, allowing storage of discrete harmonic components and/or the magnitude of the continuous frequency spectrum. Both outputs may be configured to limit the data storage to that outside user-defined thresholds. A single level may be set for the continuous spectrum, and individual levels may be set for each of the harmonic components.

Flicker : The flicker calculation requires a single voltage signal as an input and provides real time monitoring of the voltage fluctuations. The algorithm for the flicker calculations provides a short-term (P_{ST}) and long-term (P_{LT}) flicker severity level evaluation using the 50 Hz,

230V weighting curve or the 60 Hz, 120V weighting curve. The P_{ST} is a statistical evaluation of flicker over a 10-minute window and the P_{LT} is a combination of the P_{ST} values over a 2-hour window. The flicker calculation also provides an output of percent voltage fluctuation versus time, and the output can be weighted. The different weighting curves that are included are :

1. General Electric borderline of irritation curve.
2. General Electric borderline of visibility curve.
3. Commonwealth Edison (CE) curve.
4. Japanese DV10 curve.

Voltage Dip : Voltage dips are characterized by profiling the fault-induced voltage variations on a Computer Business Equipment Manufacturers Association (CBEMA) curve. Voltage excursions are saved by duration and magnitude, and/or by means of a detailed output which gives the shape of the voltage excursion. Custom voltage dip classification curves can be implemented if required.

5.5.3. Arc Furnace Module (AFM)

Since poor electrode regulation can lead to arc instability and loss of arc, effective electrode regulation is a necessary element in enhancing power quality. Power, Voltage and Current (PVI) diagrams (derived from plant specific parameters) display true characteristics of the theoretical circle diagrams. A regulator set-point would appear as a fixed dot on the PVI diagram. During operation, the true operating point would fluctuate about this dot, with the distance from the set-point being a measure of controller error.

More advanced performance measures may be implemented, such as : normalized autocorrelations computed automatically on a step change in set-point, real time FFT calculations to display the frequency content of the arc or of disturbances occurring in the bath, etc. These performance measures provide a baseline for furnace control improvements.

5.6. CBEMA AND ITIC CURVES

Computer equipment sensitivity to sags and swells can be charted in curves of acceptable sag/swell amplitude versus event duration. In the 1970s, the Computer Business Equipment Manufacturers Association (CBEMA) developed the curve of Fig. 27 utilizing historical data from mainframe computer operations, showing the range of acceptable power supply voltages for computer equipment.

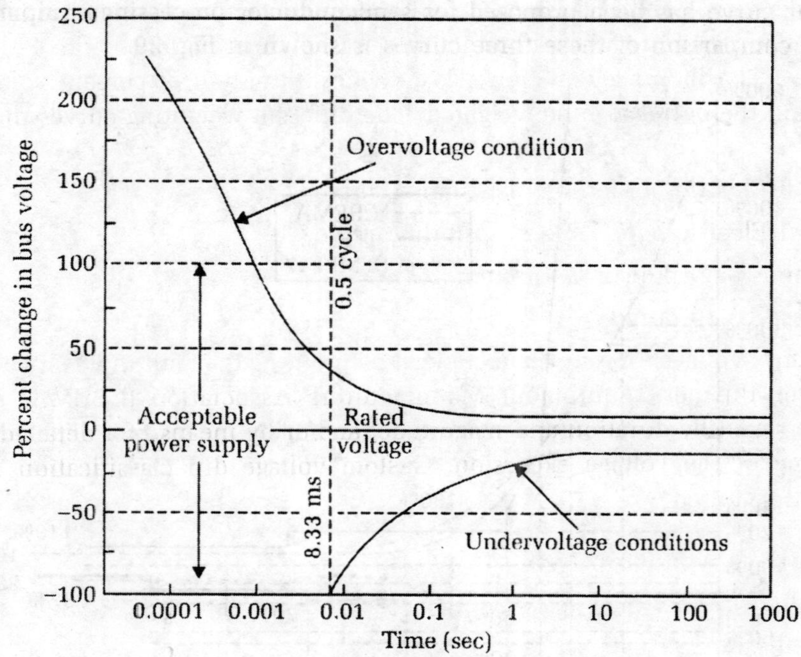

Fig. 27. *The CBEMA curve.*

The horizontal axis shows the duration of the sag or swell, and the vertical axis shows the percent change in line voltage. In addition, the IEEE has addressed sag susceptibility and the economics of sag-induced events in IEEE Std. 1346–1998. This document includes measured power quality data taken from numerous sites. In the 1990s, the Information Technology Industry Council (ITIC) curve was developed by a working group of CBEMA. In recent years, the ITIC curve (Fig. 28) has replaced the CBEMA curve in general usage for single-phase, 230-V, 50-Hz systems.

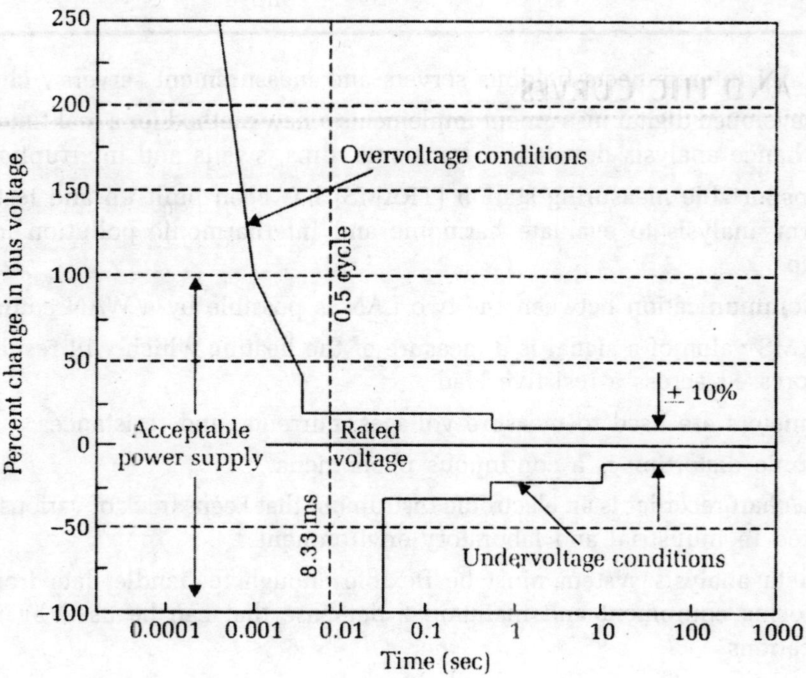

Fig. 28. *The ITIC curve.*

A similar curve has been proposed for semiconductor processing equipment : the SEMI F47 curve. A comparison of these three curves is shown in Fig. 29.

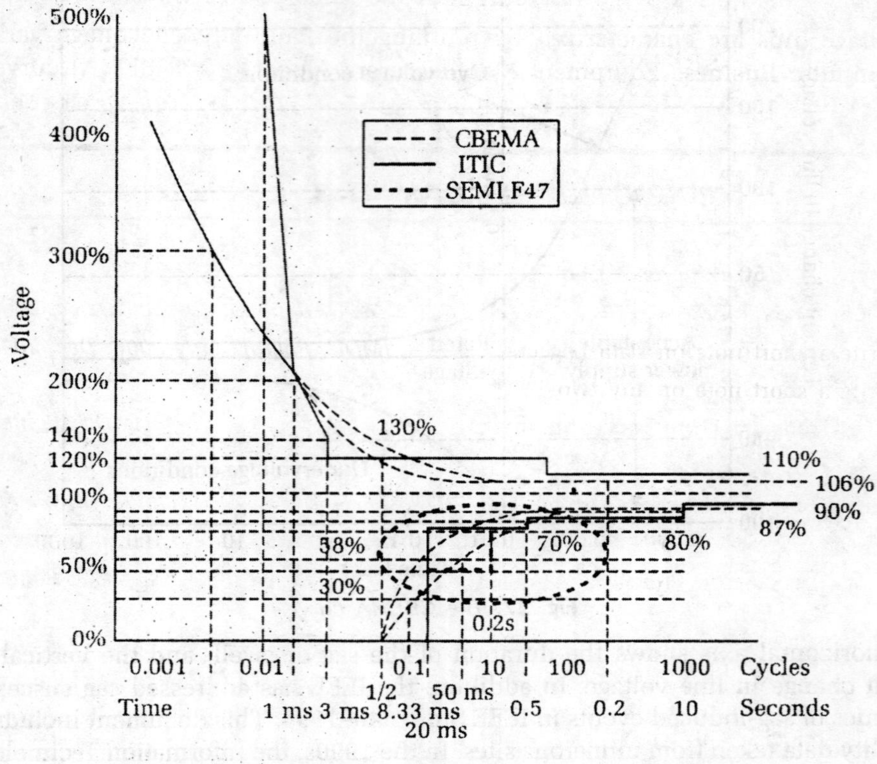

Fig. 29. *Comparison of the CBEMA, ITIC, and SEMI F47 curves.*

SUMMARY

- The LAN interconnects fieldbus servers and measurement servers / clients.
- The developed digital instrument implements a new method for a real-time non-stationary disturbance analysis devoted to transients, dips, swells and interruptions.
- A transportable measuring station (TRAMS) has been built up and tested in order to perform analysis to evaluate harmonic and interharmonic pollution for current and voltage.
- The communication between the two LAN is possible by a WAN connection.
- The RMS value of a signal is a measure of the heating which will result if the voltage is impressed across a resistive load.
- Multimeters are used to measure voltages, currents, and resistances.
- Harmonic distortion is a continuous phenomena.
- A data/chart recorder is an electronic instrument that keeps track of various measurements required in industrial and laboratory environments.
- The data analysis system must be flexible enough to handle data from a variety of monitoring equipment and maintain a database that can be used by many different applications.

- Measured voltage and current signals are corrupted by the effects of non ideal frequency responses of hardware measurement components.
- PQA calculations provide raw output at the same rate as waveform sampling.
- Voltage dips are characterized by profiling the fault-induced voltage variations on a Computer Business Equipment Manufacturers Association (CBEMA) curve.

Review Questions and Exercises

1. Explain the working of harmonic analyzer with a diagram. *(MDU, Rohtak, May 2008, Dec. 2008, May 2009)*
2. Write a short note on Data Logger. *(MDU, Rohtak, May 2008, Dec. 2008, May 2009)*
3. Write a short note on any two :
   (*i*) Transient Disturbance Analyzer *(MDU, Rohtak, May 2008)*
   (*ii*) True RMS meter *(MDU, Rohtak, May 2008, Dec. 2008, May 2009)*
   (*iii*) Chart Recorder *(MDU, Rohtak, May 2008, May 2009)*
4. Describe the operation of a transient disturbance analyzer. *(MDU, Rohtak, Dec. 2008)*
5. Explain power quality measurements. *(MDU, Rohtak, Dec. 2008)*

❑❑❑

Chapter 6

CUSTOM POWER DEVICES

6.1. INTRODUCTION

The *Custom Power* term was proposed to designate a new generation of semiconductor devices based on power electronics, designed to operate at medium and low voltage levels, and whose main objective is to improve the service quality of distribution networks. Recent advances in controllable semiconductors, micro-controllers, signal processors, and energy storage technologies have enabled the design of new devices. They can present faster responses and a more accurate setting in basic and important functions such as voltage regulation, reactive power compensation, reduction in the rate of harmonic distortion, or the limitation of short-circuit currents.

The first installation of a D-STATCOM used to compensate reactive power and to prevent the voltage 'flickers' was installed at a saw mill in British Columbia, Canada. The D-STATCOM ± 2MVAr was installed for a load of 2.6 MVA with a typical power factor of approximately 0.85.

The boom that wind generation is presenting offers the possibility of adapting these devices for supervision and correction of anomalies or distortions in the electricity network by improving the quality of supply in the wind farms. The additions of these equipments in the wind farms are having a great importance, which has an advanced static compensator for reactive power (ASVC) to improve the wave quality, and withstand voltage dips.

6.2. DYNAMIC VOLTAGE RESTORER (DVR)

Voltage magnitude is one of the major factors that determine the quality of power supply. Loads at distribution level are usually subject to frequent voltage sags due to various reasons. Voltage sags are highly undesirable for some sensitive loads, especially in high-tech industries. It is a challenging task to correct the voltage sag so that the desired load voltage magnitude can be maintained during the voltage disturbances. Dynamic voltage restorer (DVR) or distribution STATCOM (D-STATCOM) can be used to correct the voltage sag at distribution level. A DVR is a series device that generates an ac voltage and injects it in series with the supply voltage through an injection transformer to compensate the voltage sag. The injected voltage and load current determine the power injection of the DVR. On the other hand, a D-STATCOM is a shunt device that generates an ac voltage, which in turn causes a current injection into the system through a shunt transformer. The load voltage and injected current determine the power injection of the D-STATCOM.

For lower voltage sags, the load voltage magnitude can be corrected by injecting only reactive power into the system. However, for higher voltage sags, injection of active power, in addition to reactive power, is essential to correct the voltage magnitude. Note that both DVR and D-STATCOM are capable of generating or absorbing reactive power but the active power injection of the device must be provided by an external energy source or energy storage system. The response time of both DVR and D-STATCOM is very short and is limited by the power electronics devices and the voltage sag detection time. The expected response time is about 25 ms, and which is much less than some of the traditional methods of voltage correction such as tap changing transformers.

6.2.1. Voltage Sag Correction by a DVR

The schematic diagram of a typical DVR is shown in Fig. 1. The circuit on left hand side of the DVR represents the Thevenin equivalent circuit of the system. The system impedance ($Z_{th} = R_{th} + jX_{th}$) depends on the fault level of the load bus. When the system voltage (V_{th}) drops, the DVR injects a series voltage V_{DVR} through the injection transformer so that the desired load voltage magnitude V_L can be maintained. The series injected voltage of the DVR can be written as

$$V_{DVR} = V_L + Z_{th} I_L - V_{th} \quad \text{...(1)}$$

Here I_L is the load current and is given by :

$$I_L = \left(\frac{P_L + jQ_L}{V_L}\right)^* \quad \text{...(2)}$$

When V_L is considered as a reference, equation (1) can be rewritten as :

$$V_{DVR} \angle\alpha = V_L \angle 0 + Z_{th} I_L \angle(\beta - \theta) - V_{th} \angle\delta \quad \text{...(3)}$$

Here α, β and δ are the angle of V_{DVR}, Z_{th} and V_{th}, respectively, and θ is the load power factor angle ($\theta = \tan^{-1}(Q_L/P_L)$). The complex power injection of the DVR can be written as :

$$S_{DVR} = V_{DVR} I_L^* \quad \text{...(4)}$$

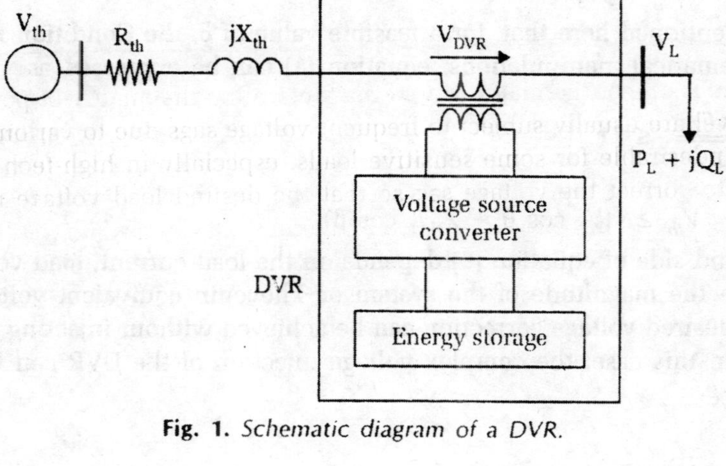

Fig. 1. *Schematic diagram of a DVR.*

It may be mentioned here that when the injected voltage V_{DVR} is kept in quadrature with I_L, no active power injection by the *DVR* is required to correct the voltage. It requires the injection of only reactive power and the *DVR* itself is capable of generating the reactive power. Note that V_{DVR} can be kept in quadrature with I_L only up to a certain value of voltage sag and beyond which the quadrature relationship cannot be maintained to correct the voltage sag. For such a case, injection of active power into the system is essential. The injected active power must be provided by the energy storage system of the *DVR*. On the other hand, when the magnitude of the *DVR* injected voltage is minimized, the desired voltage correction can be achieved with minimum apparent power injection into the system. This aspect of voltage correction is also very important because it minimizes the size of the injection transformer. The voltage correction by a *DVR* for the zero active power injection (*ZAPI*) and minimum apparent power injection (*MAPI*) cases is discussed in the following.

6.2.1.1. *Zero Active Power Injection (ZAPI)*

As mentioned earlier that, when the phase angle difference between V_{DVR} and I_L is kept at $\pi/2$, no active power injection into the system is required to correct the voltage sag. In this case, the angle α of the injected voltage V_{DVR} can be written as

$$\alpha = \frac{\pi}{2} - \theta \qquad \text{...(5)}$$

From equation (3), the angle α can also be expressed as

$$\alpha = \tan^{-1}\left[\frac{Z_{th}\, I_L \sin(\beta-\theta) - V_{th}\sin\delta}{V_L + Z_{th} I_L \cos(\beta-\theta) - V_{th}\cos\delta}\right] \qquad \text{...(6)}$$

By equating equations (5) and (6), the phase angle δ of V_{th} can be expressed as

$$\delta = \cos^{-1}\left(\frac{c_1 - c_2}{\sqrt{c_3^2 + c_4^2}}\right) - \gamma \qquad \text{...(7)}$$

where,

$$c_1 = V_L + Z_{th} I_L \cos(\beta - \theta)$$

$$c_2 = Z_{th} I_L \tan\theta \sin(\beta - \theta)$$

$$c_3 = V_{th}$$

$$c_4 = V_{th} \tan\theta$$

and

$$\gamma = \tan^{-1}(c_4/c_3)$$

It may be mentioned here that, for a feasible value of δ, the Condition must be satisfied. After some mathematical manipulations, equation (8) can be expressed as

$$\frac{c_1 - c_2}{\sqrt{c_3^2 + c_4^2}} \leq 1 \qquad \text{...(8)}$$

$$V_{th} \geq (V_L \cos\theta + Z_{th} I_L \cos\beta) \qquad \text{...(9)}$$

The right hand side of equation (9) depends on the load current, load voltage and system impedance. When the magnitude of the system or Thevenin equivalent voltage (V_{th}) satisfies equation (9), the desired voltage correction can be achieved without injecting any active power into the system. In this case, the complex voltage injection of the DVR can be obtained from equation (3) with a value of δ found from equation (7). Once the value of V_{DVRi} is known, the complex power injection of the DVR can be found from equation (4) and it will have only the reactive component.

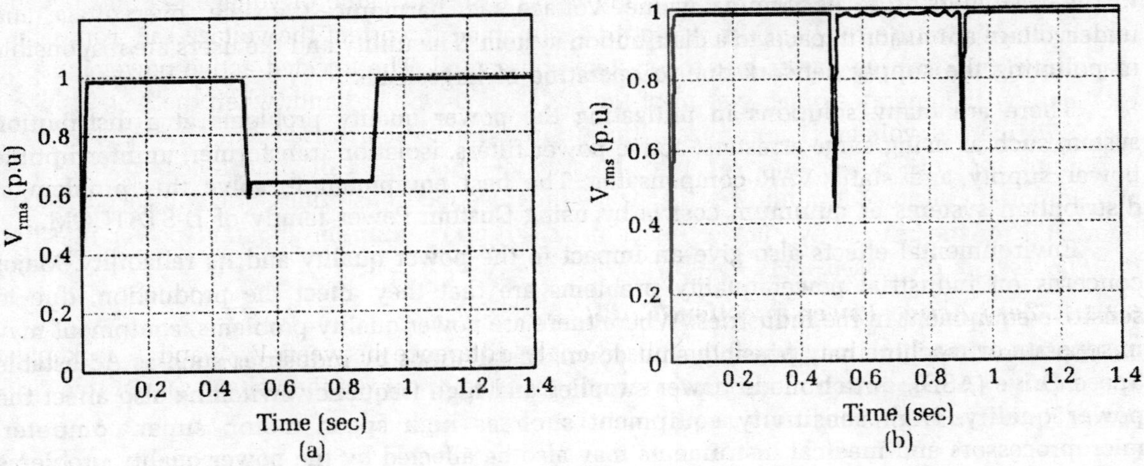

Fig. 2. *(a) Voltage response of the test system without DVR (b) Voltage response of the test system2 with DVR.*

6.2.1.2. Minimum Apparent Power Injection (MAPI)

For a given load current, the magnitude of the injected apparent power of the DVR depends on the magnitude of the injected voltage. From equation (3), the magnitude of the injected voltage can be expressed as

$$V_{DVR}^2 = V_L^2 + V_{th}^2 + Z_{th}^2 I_L^2 + 2V_L Z_{th} I_L \cos(\beta - \theta) - 2V_{th} Z_{th} I_L \cos(\beta - \theta - \delta) - 2V_L V_{th} \cos\delta \quad ...(10)$$

Thus for the minimum voltage magnitude (or minimum apparent power) injection, the condition

$$\frac{\partial V_{DVR}^2}{\partial \delta} = 0 \quad ...(11)$$

must be satisfied. Equation (11) can now be solved for the value of δ and is given by

$$\delta = \tan^{-1}\left[\frac{Z_{th} I_L \sin(\beta - \theta)}{V_L + Z_{th} I_L \cos(\beta - \theta)}\right] \quad ...(12)$$

Once the value of δ is known, the injected complex voltage and apparent power of the DVR can again be obtained from equations (3) and (4), respectively.

6.3. D-STATCOM

Electric problems always occur regardless of time and place. This may cause an impact to the electric supply thus may affect the manufacturing industry and impede the economic development in a country. The major electric problems that always occur in power systems are the power quality problems that have been discussed by the electrical engineers around the world, since problems have become a major issue due to the rapid development of sophisticated and sensitive equipment in the manufacturing and production industries.

The increased concern for power quality has resulted in measuring power quality variations, studying the characteristics of power disturbances and providing solutions to the power quality problems. In distribution systems, the power quality problems can reduce the power supplied to the customers from its nominal value. Voltage sag, harmonic, transient, overvoltage and undervoltage are major impacts to a distribution system. The utility and the users are responsible in polluting the supply network due to operating of large loads.

There are many solutions in mitigating the power quality problems at a distribution system such as using surge arresters, active power filters, isolation transformer, uninterruptible power supply and static VAR compensator. The best equipment to solve this problem at distribution systems at minimum cost is by using Custom Power family of D-STATCOM.

Environmental effects also give an impact to the power quality and its reliability. Major concerns on industrial power quality problems are that they affect the production, due to sensitive equipment in the industries. Where there are power quality problems, equipment may misoperate or machine may possibly shut down. Installations by industries such as Adjustable Speed Drive (ASD), switch mode power supplies and high frequency switching also affect the power quality. High sensitivity equipment such as high speed motor, super computer. microprocessors and medical instruments may also be affected by the power quality problems occurring in the system.

6.3.1. Basic Configuration and Operation of D-STATCOM

The D-STATCOM is a three-phase and shunt connected power electronics based device. It is connected near the load at the distribution systems. The major components of a D-STATCOM are shown in Fig. 3. It consists of a dc capacitor, three-phase inverter (IGBT, thyristor) module, ac filter, coupling transformer and a control strategy. The basic electronic block of the D-STATCOM is the voltage-sourced inverter that converts an input dc voltage into a three-phase output voltage at fundamental frequency. The D-STACOM employs an inverter to convert the DC link voltage V_{dc} on the capacitor to a voltage source of adjustable magnitude and phase. Therefore the D-STATCOM can be treated as a voltage-controlled source. The D-STATCOM can also be seen as a current-controlled source. Fig. 3 shows the inductance L and resistance R which represent the equivalent circuit elements of the stepdown transformer and the inverter will is the main component of the D-STATCOM. The voltage V_i is the effective output voltage of the D-STATCOM and δ is the power angle. The reactive power output of the D-STATCOM inductive or capacitive depending can be either on the operation mode of the D-STATCOM.

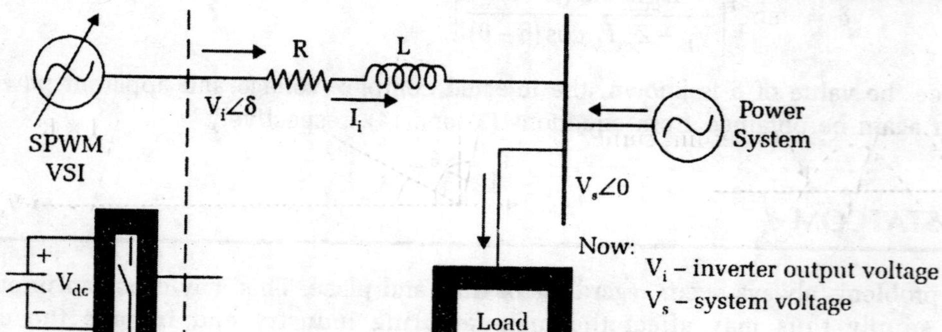

Fig. 3. *Basic building blocks of the D-STATCOM.*

Referring to Fig. 3, the construction controller of the D-STATCOM is used to operate the inverter in such a way that the phase angle between the inverter voltage and the line voltage is dynamically adjusted so that the D-STATCOM generates or absorbs the desired VAR at the point of connection. The phase of the output voltage of the thyristor-based inverter, V_i is controlled in the same way as the distribution system voltage, V_s. Fig. 4 shows the three basic operation modes of the D-STATCOM output current, I, which varies depending upon V_i. If V_i is equal to V_s, the reactive power is zero and the D-STATCOM does not generate or absorb reactive power. When V_i is greater than V_s, the D-STATCOM shows an inductive reactance connected at its terminal. The current, I, flows through the transformer reactance from the D-STATCOM to the ac system, and the device generates capacitive reactive power. If V_s is greater than V_i, the D-STATCOM shows the system as a capacitive reactance. Then the current flows from the ac system to the D-STATCOM, resulting in the device absorbing inductive reactive power.

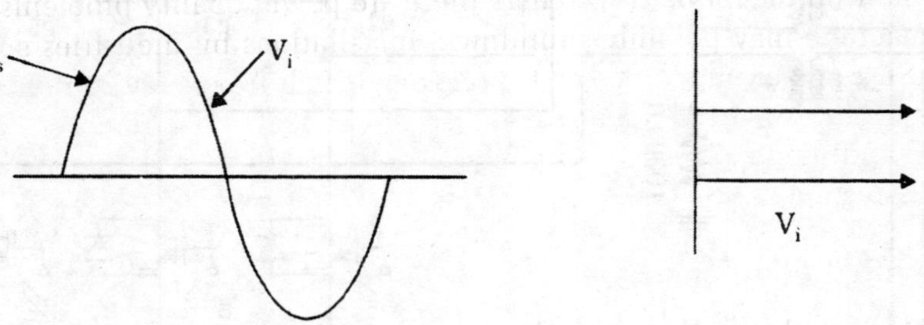

(a) No-load mode ($V_s = V_i$)

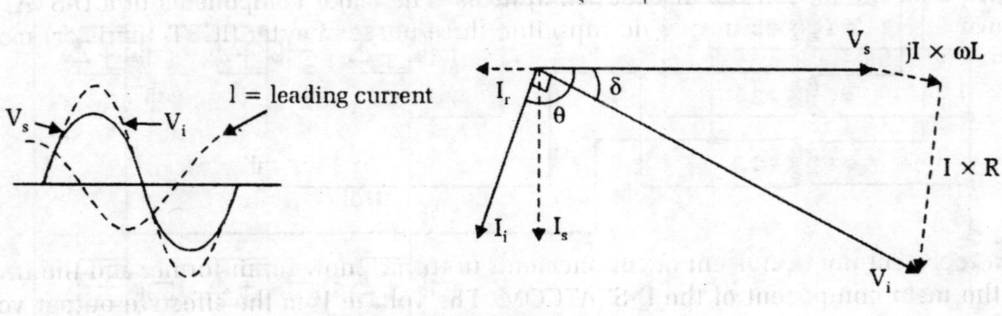

(b) Capacitive mode ($V_i > V_s$)

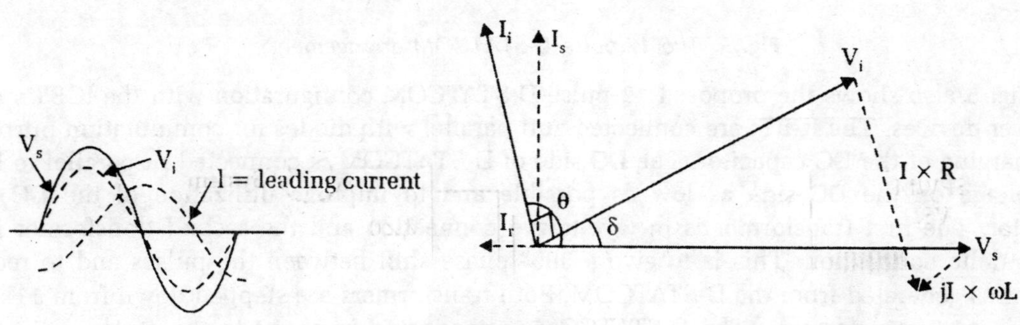

(c) Inductive mode ($V_i < V_s$)

Fig. 4. *Operation modes of D-STATCOM.*

6.3.1.1. *Construction of the 12-Pulse D-STATCOM and Control System*

Fig. 5 shows a typical 12-pulse inverter arrangement utilizing two transformers with their primaries connected in series. The first inverter is connected to the system through a *Y-Y* arrangement, whereas a *Y-Δ* connection is used for the second inverter. Each inverter operates as a 6-pulse inverter, with the *Y-Δ* inverter being delayed by 300° with respect to the *Y-Y* inverter. The current flowing into each inverter is the same, scaled by the transformer ratio, as the current being drawn from the system by the D-STATCOM. For the *Y-Δ* inverter, the current is also delayed by 300° with respect to the current of the Y-Y inverter.

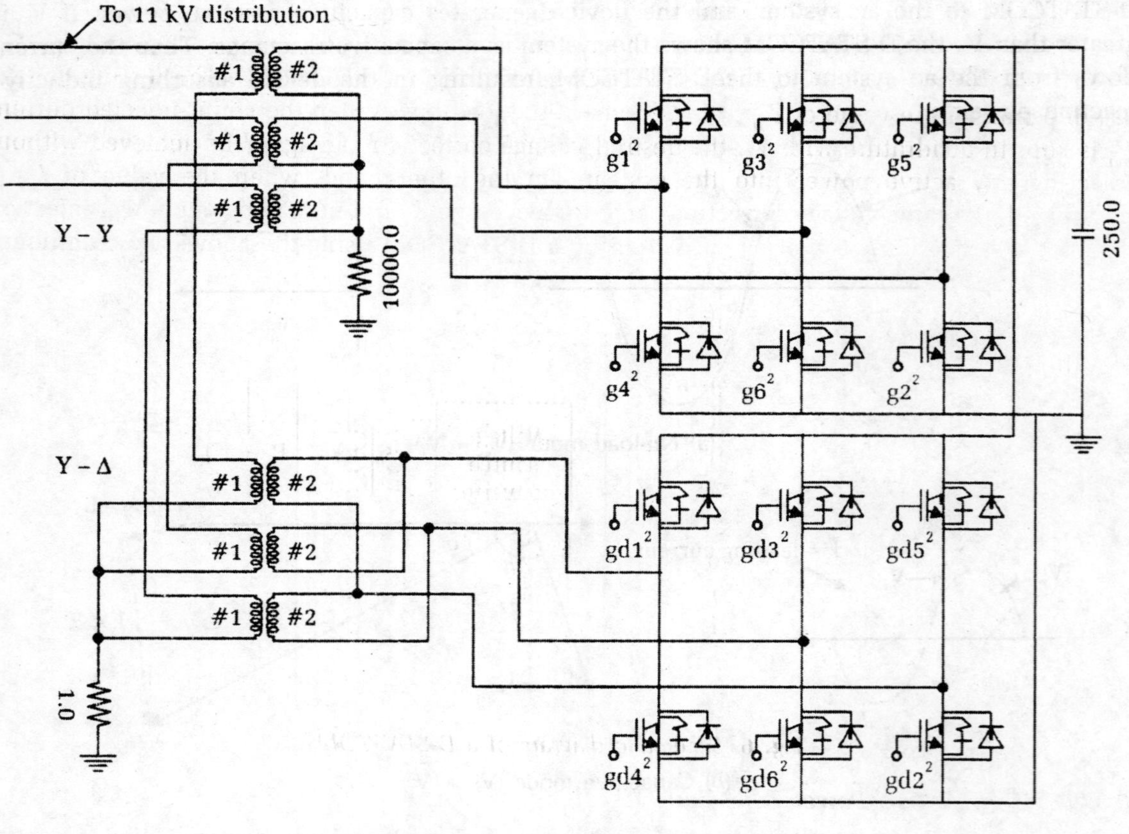

Fig. 5. *The 12-pulse D-STATCOM arrangement.*

Fig. 5 also shows the proposed 12-pulse D-STATCOM configuration with the IGBT's used as power devices. The IGBTs are connected anti parallel with diodes for commutation purposes and charging of the DC capacitor. The DC side of D-STATGOM is connected in parallel to keep the voltage on the DC side as low as possible and to improve utilization of the DC side capacitor. The first transformer is in wye-towye connection and the second transformer is in wye-to-delta connection. This is to give a 300° phase shift between the pulses and to reduce harmonics generated from the D-STATCOM. Both transformers are stepped down from 11kV to 2kV, *i.e.*, 11:2 transformers. The D-STATCOM is connected in shunt to the system.

6.3.2. Voltage Sag Correction by a D-STATCOM

The schematic diagram of a D-STATCOM is shown in Fig. 6. In this diagram, the shunt injected current I_{sh} corrects the voltage sag by adjusting the voltage drop across the system impedance Z_{th}. The value of I_{sh} can be controlled by adjusting the output voltage of the converter. The shunt injected current I_{sh} can be written as

$$I_{sh} = (I_L - I_s) = I_L - \frac{V_{th} - V_L}{Z_{th}}$$

or
$$I_{sh}\angle\eta = I_L\angle-\theta - \frac{V_{th}}{Z_{th}}\angle(\delta-\beta) + \frac{V_L}{Z_{th}}\angle-\beta \qquad ...(13)$$

The complex power injection of the D-STATCOM can be expressed as

$$S_{sh} = V_L I_{sh}^* \qquad ...(14)$$

It may be mentioned here that the effectiveness of the D-STATCOM in correcting voltage sag depends on the value of Z_{th} or fault level of the load bus. When the shunt injected current I_{sh} is kept in quadrature with V_L, the desired voltage correction can again' be achieved without injecting any active power into the system. On the other hand, when the value of I_{sh} is minimized, the same voltage correction can be achieved with minimum apparent power injection into the system. The voltage sag correction by a D-STATCOM using the above two techniques is discussed in the following.

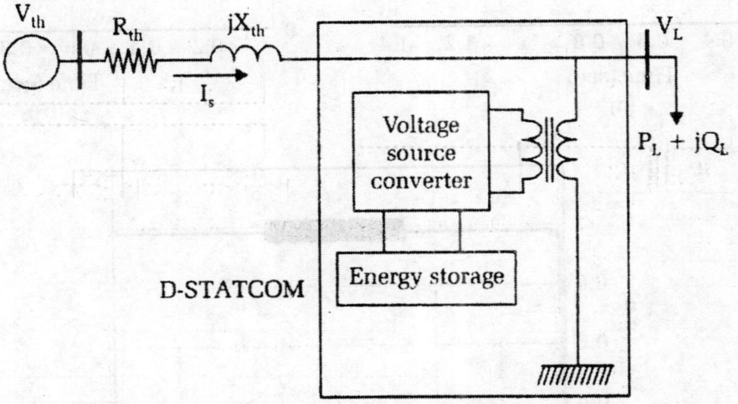

Fig. 6. *Schematic diagram of a D-STATCOM.*

6.3.2.1. Zero Active Power Injection (ZAPI)

In this case, the D-STATCOM is not injecting any active power into the system. Thus the entire load active power (P_L) must be provided by the Thevenin equivalent of the system. The active power flow through the Thevenin impedance of Fig. 6 (at load side) can be written as

$$P_L = \frac{V_{th}V_L}{Z_{th}}\cos(\beta-\delta) - \frac{V_L^2}{Z_{th}}\cos\beta \qquad ...(15)$$

From equation (15), the angle δ can be expressed as

$$\delta = \beta - \cos^{-1}\left[\frac{V_L}{V_{th}}\cos\beta + \frac{Z_{th}P_L}{V_{th}V_L}\right] \qquad ....(16)$$

For a feasible value of δ, the condition

$$\frac{V_L}{V_{th}}\cos\beta + \frac{Z_{th}P_L}{V_{th}V_L} \le 1 \qquad ...(17)$$

must be satisfied. The above constraint can be rewritten as

$$V_{th} \geq (V_L \cos \beta + Z_{th}P_L/V_L) \quad ...(18)$$

Thus, when the system voltage magnitude satisfies equation (18), the D-STATCOM can correct the voltage sag without injecting any active power into the system. For such a case, the injected complex current and apparent power of the DSTATCOM can easily be found from equations (13) and (14), respectively. Note that the injected apparent power will have only the reactive component.

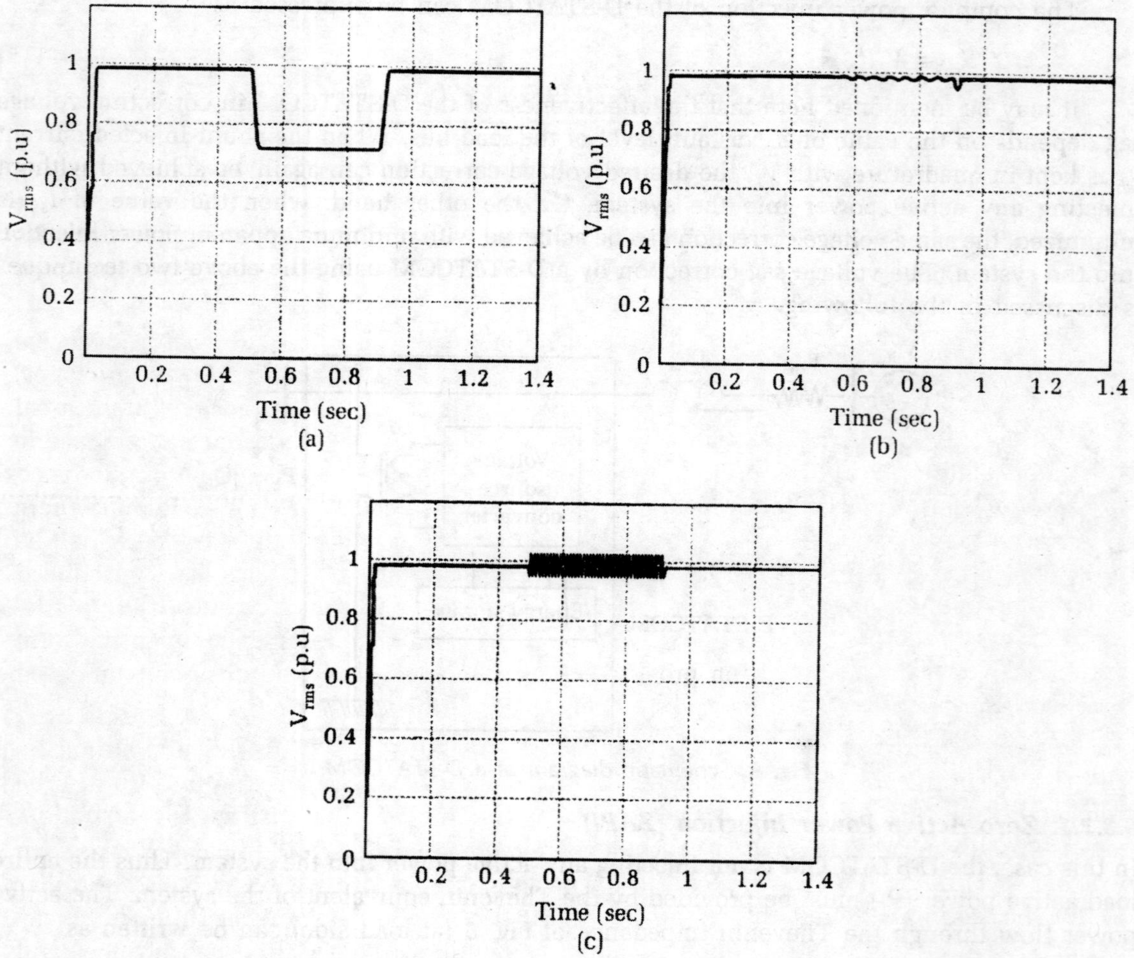

Fig. 7. *(a) Voltage response of the test system without D-STATCOM (b) Voltage response of the test system 1 with D-STATCOM; With 750μF Capacitor (c) Voltage response of the test system 1 with D-STATCOM; With 75μF Capacitor*

6.3.2.2. *Minimum Apparent Power Injection (MAPI)*

As mentioned earlier that, when the magnitude of the injected current is minimized, the D-STATCOM can correct the voltage sag with minimum apparent power injection into the system. Thus the condition of minimum apparent power injection is

$$\frac{\partial I_{sh}}{\partial \delta} = 0 \quad ...(19)$$

An analytical expression of I_{sh} can readily be obtained from equation (13), and the solution of equation (19) provides the following :

$$\delta = \tan^{-1}\left[\frac{Z_{th} I_L \sin(\beta - \theta)}{V_L + Z_{th} I_L \cos(\beta - \theta)}\right] \quad ...(20)$$

Thus for a given load, the value of 6 can easily be found from equation (20). Once the value of **δ** is known, the complex current and apparent power injection of the D-STATCOM can again be obtained from euqations (13) and (14), respectively.

6.4. UNIFIED POWER QUALITY CONDITIONER (UPQC)

The better controllability, higher efficiency, higher current carrying capability, and fast switching characteristics of static power converters are promoting major changes in controlling the power flow of transmission and distribution systems. On the other hand the nonlinear characteristics of these switching devices introduce many undesirable features such as low power factor, poor voltage regulation, zerosequence currents, imbalances, and harmonics.

Traditionally passive filters, synchronous condensers, capacitors, and phase advancers were used to improve the power quality. The undesirable features such as lower efficiency, bulkiness, fixed compensation, resonance, and electromagnetic interference of traditional compensators urged power electronics and power system engineers to develop an adjustable and dynamic solution for power quality problems. Active power filters (APF) were introduced in order to compensate reactive power, to cancel current harmonics, to correct current imbalances and to control zero-sequence currents, where as the disturbances in terminal voltages shall be compensated using dynamic voltage restorer (DVR). More recently UPQC has been introduced as a one shot solution to improve power quality. For this a multilevel converter using diode-clamped inverters that can handle higher voltages with extra degrees of freedom in the form of redundant voltage states has been proposed. A new topology and control circuit proposed in has been successfully tested for various operating conditions.

UPQC uses a control circuit without reference calculation. Complicated control structures of UPQC have been replaced by a simple control technique. Voltage interruption can also be eliminated by the use of a unified power quality conditioner with distributed generation. Recent research shows that AI based controllers are very promising in the field power system and power electronics. The conventional PI controller is replaced by a fuzzy logic controller [FLC] for the determination of the reference current in a shunt active power filter. T-S fuzzy model is used to predict future harmonic compensating current in an APF system. Successful application of artificial neural network (ANN) controllers for the implementation of APF system has been reported and many disadvantages of conventional controllers could be eliminated by the use of ANN controllers.

6.4.1. Principle of UPQC

UPQC is one of the custom power devices used at the electrical power distribution systems to improve the power quality of distribution system customers. UPQC could be used to cancel current harmonics, to compensate reactive power, to eliminate voltage harmonics, to improve voltage regulation, to correct voltage and current imbalances, to correct voltage sag or swell and to avoid voltage interruptions.

A UPQC consists of both shunt and series compensators. A shunt compensator is used to cancel the disturbances in current whereas series compensator is used to cancel disturbances in voltage. Shunt compensator could be connected to the left or right of the series compensator. Ideally, shunt compensator injects current to achieve purely balanced sinusoidal source currents in phase with the supply voltages at rated magnitude and frequency. On the other hand series compensation is used to inject voltage to maintain terminal voltage at rated magnitude and frequency.

6.4.2. Control Circuit of UPQC

Figure 8 shows the schematic diagram of the UPQC system. Voltage source inverters are used for shunt and series compensation. One may note that both voltage source inverters are supplied from a common dc link capacitor. One of the voltage source inverters is connected in parallel with the a.c. system while the other one is connected in series with the a.c. system through injection transformers. The inverter connected in parallel, together with its control circuit forms the shunt compensation circuit. On the other hand the inverter connected in series with appropriate control circuit forms the series compensation circuit. For the successful operation of the UPQC, the dc capacitor voltage should be at least 150 % of maximum line-line supply voltage. To regulate the capacitor voltage constant, either a PI controller or a fuzzy controller could be used. Thus the control structure of UPQC has been divided into shunt compensator and series compensator control circuits.

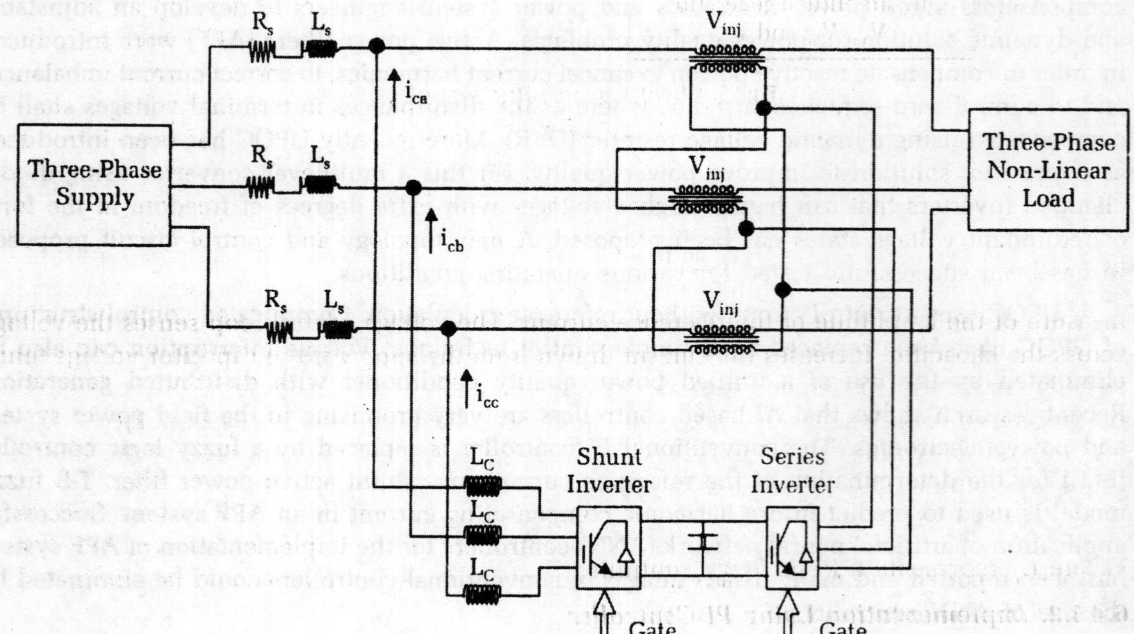

Fig. 8. *Schematic diagram of a UPQC.*

Using Park's transformation actual receiving end voltages (V_{ra}, V_{rb}, and V_{rc}) may be decomposed into stationary axes as follows :

$$V_d = \sqrt{\frac{2}{3}}\left(V_{ra}\cos\theta + V_{rb}\cos(\theta - 120°) + V_{rc}\cos(\theta - 240°)\right)$$

$$V_q = -\sqrt{\frac{2}{3}}\left(V_{ra}\sin\theta + V_{rb}\sin(\theta - 120°) + V_{rc}\sin(\theta - 240°)\right)$$

$$V_0 = \sqrt{\frac{2}{3}}\left(\frac{1}{\sqrt{2}}V_{ra} + \frac{1}{\sqrt{2}}V_{rb} + \frac{1}{\sqrt{2}}V_{rc}\right)$$

6.5. UNIFIED POWER QUALITY CONDITIONER BASED ON CURRENT SOURCE CONVERTER TOPOLOGY

Nonlinear devices, like power electronics converters, inject harmonic currents in the AC system and increase overall reactive power demanded by the equivalent load. Also, the number of sensitive loads that require ideal sinusoidal supply voltages for their proper operation has increased. In order to keep power quality under limits proposed by standards, it is necessary to include some sort of compensation. A solution based on flexible AC transmission systems (FACTS) have appeared in recent times. FACTS converters have been modified to serve in distribution network and, through a modification of a unified power flow controller (UPFC) the unified power quality conditioner (UPQC) was presented during 1998. Such solution can compensate for different power quality phenomena, such as : sags, swells, voltage imbalance, flicker, harmonics and reactive currents.

UPQC usually consists of two voltage-source converters sharing the same capacitive DC link. One of the converters is an active rectifier (AR) while other is a series filter (SF) with a LC ripple filter and transformer isolation from power supply network. Also, at the point of the load connection, passive filter banks are connected.

Voltage-source topology of UPQC has its drawbacks, mainly inside the SF, such as a rather ... (LC filter), output voltage and current protection problems. Also.

6.4.3. Shunt Compensation Control Circuit

A current controlled VSI connected in parallel to the source through booster inductors functions as the shunt compensator. (Fig. 9) shows the block diagram of the control circuit of the shunt compensator. The control circuit consists of a voltage control loop and two current control loops. The ideal requirements of the shunt compensator circuit are (*i*) to maintain the capacitor voltage at a constant steady value and (*ii*) to maintain the source current purely sinusoidal in nature and in phase with the supply voltage.

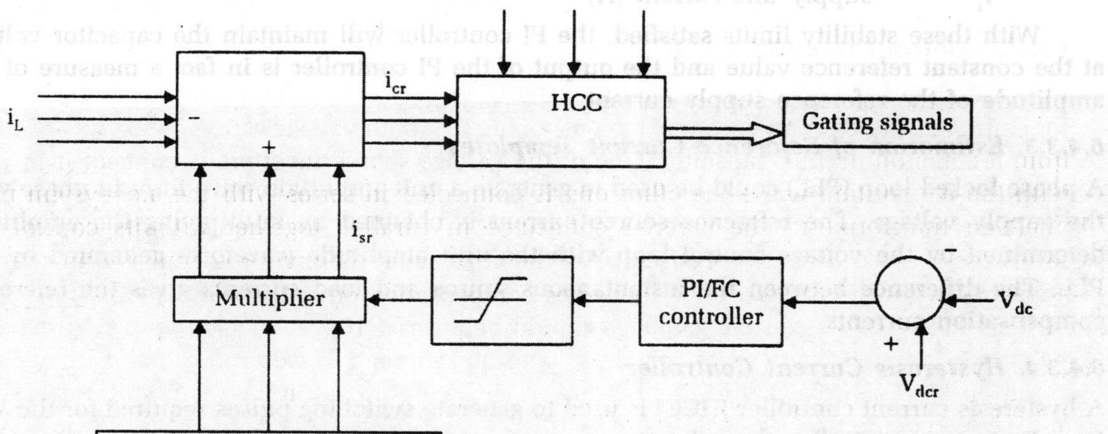

Fig. 9. *Shunt compensator control circuit.*

6.4.3.1. Voltage Control Loop

The voltage control loop is used to determine the amplitude of the reference source current. Under steady state conditions the d.c. load removes energy from capacitor at constant average rate and the capacitor voltage can be maintained constant only if the incoming power from the a.c. side is equal to the output power demand. Thus the variation in capacitor voltage is a measure of the amplitude of the reference current. The voltage control loop senses the voltage across the capacitor, increases the current drawn from the supply if the capacitor voltage tends to decrease from the reference value and decreases the current drawn from the ac side when the capacitor voltage tends to increase. Thus the voltage control loop monitors the output voltage and determines the amplitude of the source current. The amplitude of the reference source current could be evaluated as follows :

$$I_{max} = G_C\,(V_{dcr} - V_{dc})$$

Where G_c represents a PI or fuzzy controller.

6.4.3.2. Implementation Using PI Controller

According to stability criteria, the following relations may be used to determine the value of K_p and K_i of the PI controller.

$$I_s \le \frac{C_D V_{dcr}}{3K_P L_S}$$

$$I_s \le \frac{K_p V\cos\phi}{2R_s K_p + L_s K_i}$$

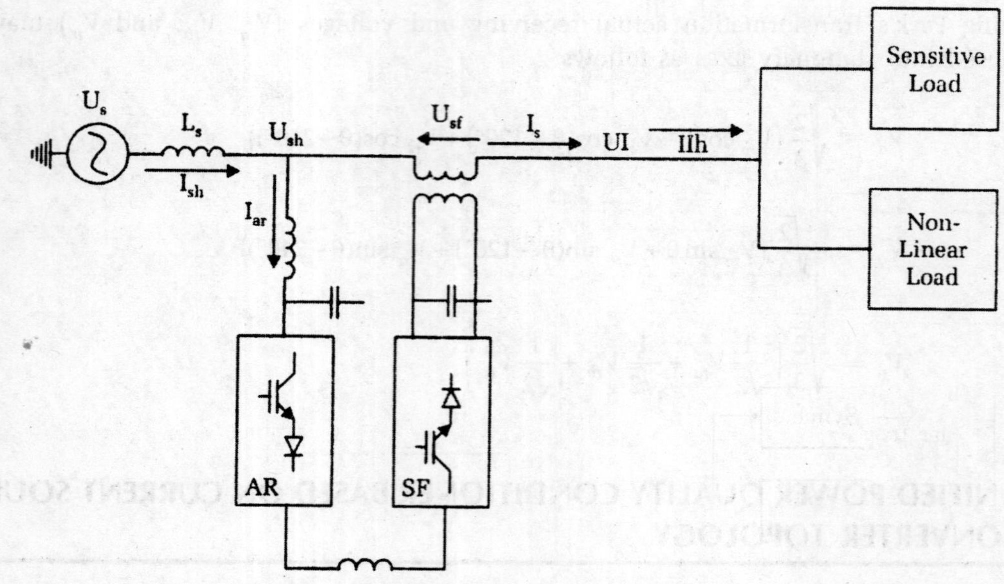

Fig. 11. *UPQC topology using three-phase current-source converters.*

6.5.1. Converter Configuration

UPQC consists of the :

- Series filter (SF) that compensates supply voltage harmonics, flicker, voltage sags/swells and supply voltage unbalance. Its control equation is :

$$U_{sf} = U_{copm}$$

Where U_{comp} is compensation voltage needed to remove supply voltage imperfections. It consists of two components—one extracted by dq theory for harmonics detection, and another determined by the Fortes cue transform, for unbalance compensation.

- Active rectifier (AR) for real power transfer to/from common DC bus and for DC bus current control with a unity power factor, as well as shunt harmonic filtering and/or reactive power compensation. DC bus current is maintained constant using three-phase buck converter, controlled in such a way that it is insensitive to supply voltage imperfections. Control equation of the active rectifier is :

$$I_{ar} = I_{ar1} + I_{pf}$$

Where I_{ar1} is the first harmonic current at the rectifier input, commanded by the DC current control loop and I_{pf} is the current component for harmonic filtering and power factor correction.

6.5.2. Control of Current Source Converters

Control structure for current source converter based UPQC is presented in Fig. 12. Voltage reference signals are compared to actual output voltages. Their difference (voltage error) is fed to hysteresis comparators forming the inverter bi-level switching signals SB_{SF}. Rectified and filtered voltage error forms DC bus current reference. DC bus current is controlled by simple PI controller, which output is synchronized with individual supply phase voltages, forming the first harmonic current reference component. These references are compared to the actual currents

amplitude of the reference supply current.

6.4.3.3. Estimation of Reference Current Templates

A phase locked loop (PLL) could be used to generate a unit amplitude wave form in phase with the supply voltage. The reference source current is obtained by multiplying the amplitude determined by the voltage control loop with the unit amplitude waveform generated by the PLL. The difference between the instantaneous source and load currents gives the reference compensation currents.

6.4.3.4. Hysteresis Current Controller

A hysteresis current controller [HCC] is used to generate switching pulses required for the VSI. In a hysteresis controller the reference compensation current is compared with the actual current that is being injected by the compensation circuit. A positive pulse is produced if the actual current tends to decrease below the lower hysteresis limit, while a negative pulse is produced if the current exceeds the upper hysteresis limit. Thus in a hysteresis current controller the actual compensation current is forced to stay within a particular hysteresis band.

6.4.4. Series Compensation Control Circuit

A DVR based on a pulse-width modulated (PWM) VSI, which is capable of generating or absorbing real and reactive power independently at its ac output terminals is used as the series compensator. The series compensator injects three single-phase ac voltages in series and in synchronism with the upstream voltages in the distribution system. The power circuit could be divided into two main parts : three-phase VSI, three single-phase injection transformers. The three single-phase transformers are connected to the distribution system with a star/open winding. The star/open winding allows the injection of positive, negative and zero sequence voltages. The windings on the inverter side are connected in delta through inductors to provide high voltage. The block diagram of a series compensator is shown in (Fig. 10).

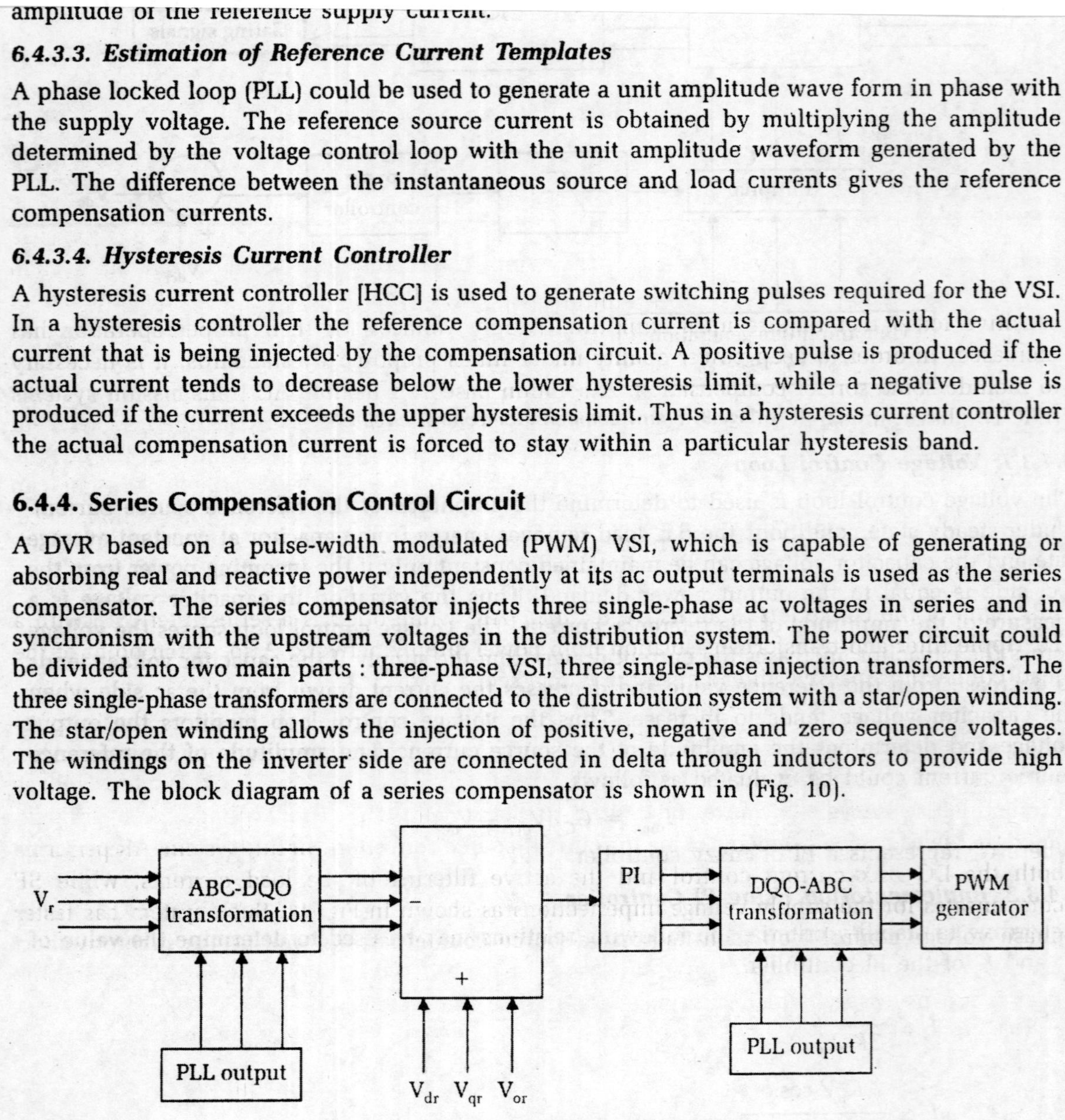

Fig. 10. *Series compensator control circuit.*

at the converters input and their difference is fed to the hysteresis controller which output is then AR bi-level switching signals SB_{AR}.

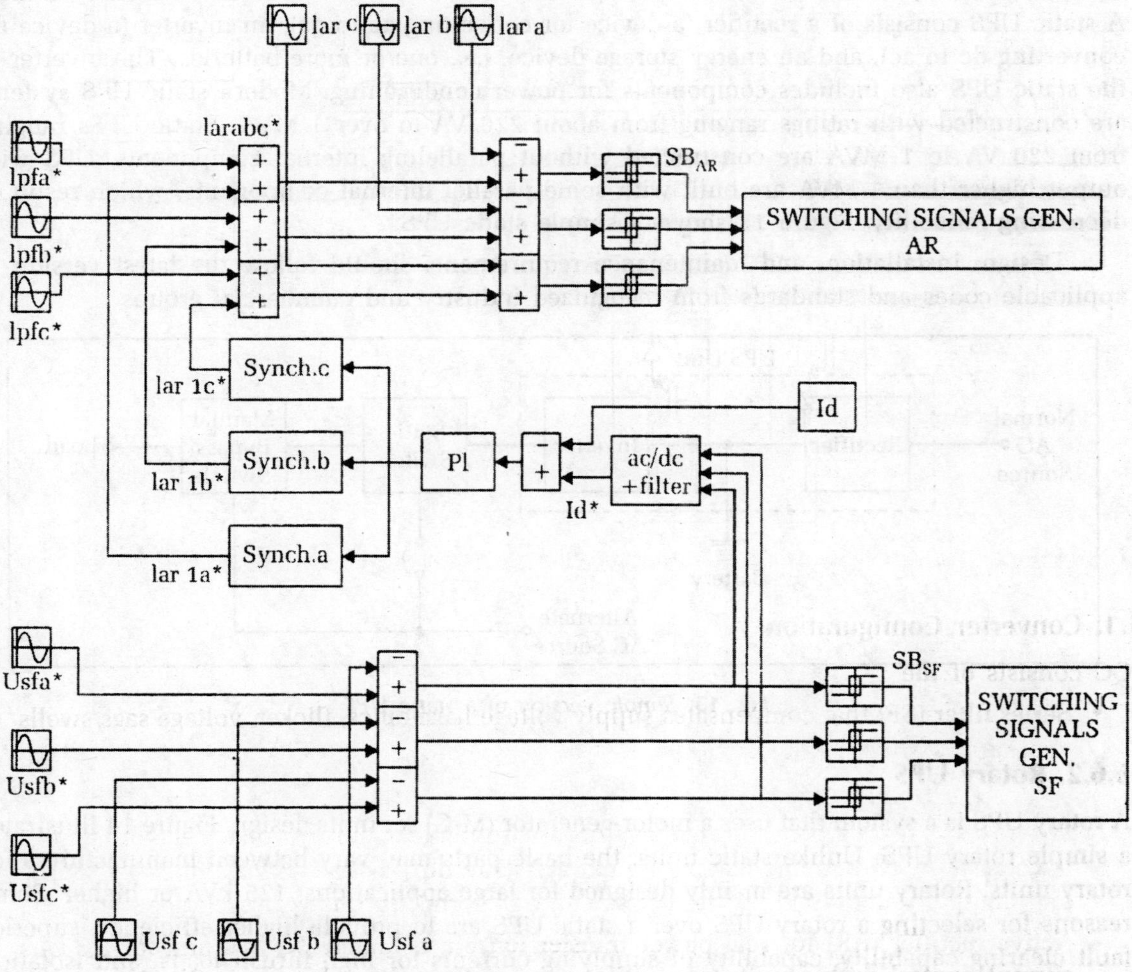

Fig. 12. *UPQC control structure.*

6.6. PRINCIPLES AND CONFIGURATIONS OF UPS

An UPS system is an alternate or backup source of power with the electric utility company being the primary source. The UPS provides protection of load against line frequency variations, elimination of power line noise and voltage transients, voltage regulation, and uninterruptible power for critical loads during failures of normal utility source. An UPS can be considered a source of standby power or emergency power depending on the nature of the critical loads. The amount of power that the UPS must supply also depends on these specific needs. These needs can include emergency lighting for evacuation, emergency perimeter lighting for security, orderly shut down of manufacturing or computer operations, continued operation of life support or critical medical equipment, safe operation of equipment during sags and brownouts, and a combination of the preceding needs.

6.6.1. Static UPS

A static UPS is a solid-state system relying solely on battery power as an emergency source. A static UPS consists of a rectifier (a device for converting ac to dc), an inverter (a device for converting dc to ac), and an energy storage device, *i.e.*, one or more batteries. The inverter in the static UPS also includes components for power conditioning. Modern static UPS systems are constructed with ratings ranging from about 220 VA to over 1 MVA. Static UPSs ranging from 220 VA to 1 MVA are constructed without paralleling internal components. UPS with output higher than 1 MVA are built with some parallel internal components, which result in decreasing reliability. Figure 13 shows a simple static UPS.

Design, installation, and maintenance requirements should follow the latest version of applicable codes and standards from recognized industry and commercial groups.

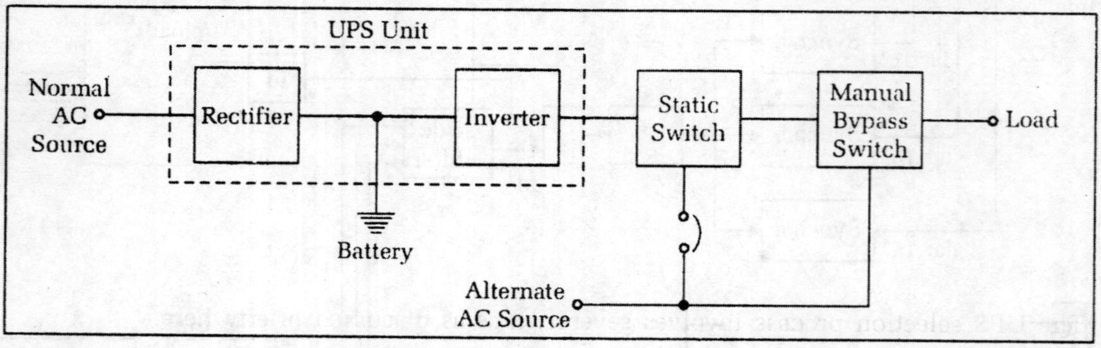

Fig. 13. *Simple version of a static UPS.*

6.6.2. Rotary UPS

A rotary UPS is a system that uses a motor-generator (M-G) set in its design. Figure 14 illustrates a simple rotary UPS. Unlike static units, the basic parts may vary between manufacturers for rotary units. Rotary units are mainly designed for large applications, 125 kVA or higher. Some reasons for selecting a rotary UPS over a static UPS are to provide higher efficiency, superior fault clearing capability, capability of supplying currents for high inrush loads, and isolation from harmonic distortion generated by non-linear loads in the line. Rotary UPS bearings must be replaced periodically. Although this might make reliability between the two types debatable, bearing failure is highly predictable with stringent routine testing. Rotary units produce more heat than do static units due to their M-G sets. They are more costly for small capacities but become competitive with static units around 300 kVA. Rotary units provide complete electrical isolation where the static UPS is limited by the static switch. Extremely high voltages or rapidly rising voltages can pass through the static switch and damage critical loads.

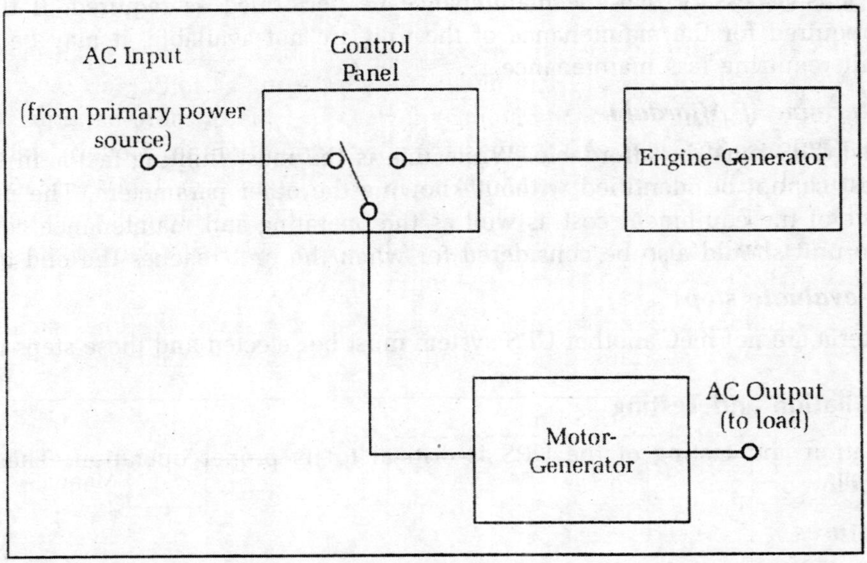

Fig. 14. *Rotary UPS (shown with primary power on).*

6.6.3. Design Criteria and Selection

The UPS selection process involves several steps as discussed briefly here.

6.6.3.1. Determine Need

Prior to selecting the UPS it is necessary to determine the need. The types of loads may determine whether local, state, or federal laws mandate the incorporation of an UPS. An UPS may be needed for a variety of purposes such as lighting, startup power, transportation, mechanical utility systems, heating, refrigeration, production, fire protection, space conditioning, data processing, communication, life support, or signal circuits. Some facilities need an UPS for more than one purpose. It is important to determine the acceptable delay between loss of primary power and availability of UPS power, the length of time that emergency or backup power is required, and the criticality of the load that the UPS must bear. All of these factors play into the sizing of the UPS and the selection of the type of the UPS.

6.6.3.2. Determine Safety

It must be determined if the safety of the selected UPS is acceptable. The UPS may have safety issues such as hydrogen accumulation from batteries, or noise pollution from solid-state equipment or rotating equipment. These issues may be addressed through proper precautions or may require a selection of a different UPS.

6.6.3.3. Determine Availability

The availability of the selected UPS must be acceptable. The criticality of the loads will determine the necessary availability of the UPS. The availability of an UPS may be improved by using different configurations to provide redundancy. It should be noted that the C4ISR facilities require a reliability level of 99.9999 percent.

6.6.3.4. Determine Maintainability

The selected UPS must be maintainable. Maintenance of the unit is important in assuring the unit's availability. If the unit is not properly cared for, the unit will be more likely to fail.

Therefore, it is necessary that the maintenance be performed as required. If the skills and resources required for the maintenance of the unit are not available, it may be necessary to select a unit requiring less maintenance.

6.6.3.5 Determine if Affordable

The selected UPS must be affordable. While this is the most limiting factor in the selection process, cost cannot be identified without knowing the other parameters. The pricing of the unit consists of the equipment cost as well as the operating and maintenance costs. Disposal costs of the unit should also be considered for when the unit reaches the end of its life.

6.6.3.6. Re-evaluate steps

If these criteria are not met, another UPS system must be selected and these steps re-evaluated.

6.6.4. Installation and Testing

The installation and testing of the UPS is critical to its proper operation. These items are discussed below :

6.6.4.1. Features

The UPS shall be installed with all necessary features. Features such as alarms, indicators, control devices, and protective devices are installed to assist in the safe operation of the unit. Power and control components such as meters, indicating lights, control switches, push buttons, and potentiometers are typically located in a nearby cabinet. Batteries are typically installed on battery racks. The design of the racks varies based on the available space and number of batteries.

6.6.4.2. Location

The UPS shall be installed on a level surface with sufficient clearance to allow for ventilation and access to maintenance panels. Static UPSs require environments with a controlled atmosphere where the temperature, humidity, and dust levels are carefully maintained. The batteries of the UPS require ventilation of the room to prevent hydrogen buildup. Rotary UPSs are suitable for placement in industrial environments.

6.6.4.3. Protection

The UPS power distribution system shall be designed to provide short circuit protection, isolate branch faults, and isolate critical loads from sources of harmonics, surges, and spikes. This is achieved using panel boards, circuit breakers, and fuses. The UPS system is grounded to ensure the safety of the operating personnel. Shielding of the control cables shall be achieved by running power cables in bonded metal enclosures separately from the control cable's enclosures.

6.6.4.4. Testing and Startup

Testing and startup shall be performed to ensure the component's safe and proper operation once energized. Acceptance testing should be performed on all equipment. Testing records on test forms should be kept for comparison to later routine maintenance tests. The possible failures of the equipment drawn out from the test results should be discussed and corrective action implemented. Test equipment used should be in accordance with the manufacturer's recommendation.

6.6.5. Maintenance

Maintenance of the UPS consists of preventive and corrective maintenance. Preventive maintenance consists of a scheduled list of activities. Performing these activities keeps the UPS in good working order and helps to prevent failures. Corrective maintenance is performed as a result of a failure.

6.6.6. Comparison between Rotary UPS and Static UPS

Since these are different types of UPS they have many differences as listed in Table 1.

Table 1 : Comparison between rotary UPS and static UPS

		Rotary UPS	***Static UPS***
1	Type	Mechanical	Electronic
2	Price (approx)	2	1
3	Weight	Very Heavy	Heavy
4	Structure foundation	Very strong	Strong
5	Noise	Very Noisy	Noisy
6	Installed location	Isolated	Normal
7	Air-Condition	Not Required	Prefered
8	Capacity per unit	Very Large	Max 600 KVA
9	Parallel function	Unknown	Up to 4
10	Redundancy	Yes	Yes
11	Reliability	High	Medium
12	Installation ratio	1	1000
13	O/P voltage stability	Low	Very High
14	O/P frequency stability	Low	Very High
15	Overload capability	High	Medium
16	Ability to take severe load unbalance	Unknown	Yes
17	Local support	Extremely Few	Many
18	User self support	Very Difficult	Possible
19	Supplier keep parts	Very Few	Many
20	Parts Weights	Very Heavy	Very Light
21	Common Parts	Very Few	Many
22	Part Prices	Very High	Low to Medium
23	Service Cost	Very High	Low to Medium
24	Local Specialist	Almost none	Many
25	Local Repair Skill	Very Few	Many
26	Support Dependency	Manufacturer	Local Distributor
27	Remote Control and/or	Unknown	Yes (optional)

SUMMARY

- The *Custom Power* term was proposed to designate a new generation of semiconductor devices based on power electronics.
- A DVR is a series device that generates an ac voltage and injects it in series with the supply voltage through an injection transformer to compensate the voltage sag.
- A D-STATCOM is a shunt device that generates an ac voltage, which in turn causes a current injection into the system through a shunt transformer.
- The magnitude of the injected apparent power of the DVR depends on the magnitude of the injected voltage.
- The better controllability, higher efficiency, higher current carrying capability, and fast switching characteristics of static power converters are promoting major changes in controlling the power flow of transmission and distribution systems.
- UPQC could be used to cancel current harmonics, to compensate reactive power, to eliminate voltage harmonics, to improve voltage regulation, to correct voltage and current imbalances, to correct voltage sag or swell and to avoid voltage interruptions.
- A UPQC consists of both shunt and series compensators.
- A shunt compensator is used to cancel the disturbances in current whereas series compensator is used to cancel disturbances in voltage.
- A static UPS is a solid-state system relying solely on battery power as an emergency source.
- A rotary UPS is a system that uses a motor-generator (M-G) set in its design.
- Maintenance of the UPS consists of preventive and corrective maintenance.

Review Questions And Exercises

1. What are static uninterruptible power supplies ? Discuss the classification and types of UPS. *(UPTU, 2007)*
2. What are series and shunt compensators ? Compare their role in power quality improvement. *(UPTU, 2007)*
3. Write a short note on Unified power quality conditioner (UPQC). *(UPTU, 2007)*
4. Explain the method of voltage sag compensation using a DVR, in detail.
5. Write a short note on D-STATCOM.

□□□

GLOSSARY

- ***Active Filter :*** Any of a number of sophisticated power electronic devices for eliminating harmonic distortion.
- ***CBEMA Curve :*** A set of curves representing the withstand capabilities of computers in terms of the magnitude and duration of the voltage disturbance. Developed by the Computer Business Equipment Manufacturers Association (CBEMA), it has become a de facto standard for measuring the performance of all types of equipment and power systems, and is commonly referred to by this name.
- ***Common Mode Voltage :*** The noise voltage that appears equally from current-carrying conductor to ground.
- ***Coupling :*** Circuit element or elements, or network, that may be considered common to the input mesh and the output mesh and through which energy may be transferred from one to another.
- ***Crest Facto :*** A value reported by many power quality monitoring instruments representing the ratio of the crest value of the measured waveform to the rms of the fundamental. For example, the crest factor of a sinusoidal wave is 1.414.
- ***Critical Load :*** Devices and equipment whose failure to operate satisfactorily jeopardizes the health or safety of personnel, and/or results in loss of function, financial loss, or damage to property deemed critical by the user.
- ***Current Distortion :*** Distortion in the ac line current. See Distortion.
- ***Differential Mode Voltage :*** The voltage between any two of a specified set of active conductors.
- ***Dip :*** See Sag.
- ***Distortion :*** Any deviation from the normal sine wave for an ac quantity.
- ***Dropout :*** A loss of equipment operation (discrete data signals) due to noise, sag, or interruption.
- ***Dropout Voltage :*** The voltage at which a device will release to its de-energized position (for this document, the voltage at which a device fails to operate).
- ***Electromagnetic Compatibility :*** The ability of a device, equipment or system to function satisfactorily in its electromagnetic environment without introducing intolerable electromagnetic disturbances to anything in that environment.
- ***Equipment Grounding Conductor :*** The conductor used to connect the non-current carrying parts of conduits, raceways, and equipment enclosures to the grounded

conductor (neutral) and the grounding electrode at the service equipment (main panel) or secondary of a separately derived system (*e.g.*, isolation transformer).

- ***Failure Mode :*** The effect by which failure is observed.
- ***Fast Tripping :*** Refers to the common utility protective relaying practice in which the circuit breaker or line recloser operates faster than a fuse can blow. Also called fuse saving. Effective for clearing transient faults without a sustained interruption, but is somewhat controversial because industrial loads are subjected to a momentary or temporary interruption.
- ***Fault :*** Generally refers to a short circuit on the power system.
- ***Fault, Transient :*** A short circuit on the power system usually induced by lightning, tree branches, or animals which can be cleared by momentarily interrupting the current.
- ***Flicker :*** Impression of unsteadiness of visual sensation induced by a light stimulus whose luminance or spectral distribution fluctuates with time.
- ***Frequency Deviation :*** An increase or decrease in the power frequency. The duration of a frequency deviation can be from several cycles to several hours.
- ***Frequency Response :*** In power quality usage, generally refers to the variation of impedance of the system, or a metering transducer, as a function of frequency.
- ***Fundamental (Component) :*** The component of order 1 (50 to 60 Hz) of the Fourier series of a periodic quantity.
- ***Ground :*** A conducting connection, whether intentional or accidental, by which an electric circuit or equipment is connected to the earth, or to some conducting body of relatively large extent that serves in place of the earth. Note : It is used for establishing and maintaining the potential of the earth (or of the conducting body) or approximately that potential, on conductors connected to it, and for conducting ground currents to and from earth (or the conducting body).
- ***Ground Electrode*** : It is a conductor or group of conductors in intimate contact with the earth for the purpose of providing a connection with the ground.
- ***Ground Grid :*** A system of interconnected bare conductors arranged in a pattern over a specified area and on or buried below the surface of the earth. The primary purpose of the ground grid is to provide safety for workmen by limiting potential differences within its perimeter to safe levels in case of high currents which could flow if the circuit being worked became energized for any reason or if an adjacent energized circuit faulted. Metallic surface mats and gratings are sometimes utilized for the same purpose. This is not necessarily the same as a Signal Reference Grid.
- ***Ground Loop :*** A potentially detrimental loop formed when two or more points in an electrical system that are nominally at ground potential are connected by a conducting path such that either or both points are not at the same ground potential.
- ***Ground Window :*** It is the area through which all grounding conductors including metallic raceways enter a specific area. It is often used in communications systems through which the building grounding system is connected to an area that would otherwise have no grounding connection.
- ***Harmonic (component) :*** A component of order greater than one of the Fourier series of a periodic quantity.

- ***Harmonic Content :*** The quantity obtained by subtracting the fundamental component from an alternating quantity.
- ***Harmonic Distortion :*** Periodic distortion of the sine wave. See Distortion and Total Harmonic Distortion (THD).
- ***Harmonic Filter :*** On power systems, a device for filtering one or more harmonics from the power system. Most are passive combinations of inductance, capacitance, and resistance. Newer technologies include active filters that can also address reactive power needs.
- ***Harmonic Number :*** The integral number given by the ratio of the frequency of a harmonic to the fundamental frequency.
- ***Harmonic Resonance :*** A condition in which the power system is resonating near one of the major harmonics being produced by nonlinear elements in the system, thus exacerbating the harmonic distortion.
- ***Impulse :*** A pulse that, for a given application, approximates a unit pulse or a Dirac function. When used in relation to the monitoring of power quality, it is preferred to use the term impulsive transient in place of impulse.
- ***Impulsive transient :*** A sudden non-power frequency change in the steady state condition of voltage or current that is unidirectional in polarity (primarily either positive or negative).
- ***Instantaneous :*** When used to quantify the duration of a short duration variation as a modifier, refers to a time range from one-half cycle to 30 cycles of the power frequency.
- ***Instantaneous Reclosing :*** A term commonly applied to reclosing of a utility breaker as quickly as possible after interrupting fault current. Typical times are 18-30 cycles.
- ***Interharmonic (component) :*** A frequency component of a periodic quantity that is not an integer multiple of the frequency at which the supply system is designed to operate (*e.g.*, 50 Hz or 60 Hz).
- ***Interruption,* Momentary (electric power systems) :** An interruption of duration limited to the period required to restore service by automatic or supervisory-controlled switching operations or by manual switching at locations where an operator is immediately available. Note : Such switching operations must be completed in a specified time not to exceed 5 minutes.
- ***Interruption, Momentary (power quality monitoring) :*** A type of short duration variation. The complete loss of voltage (<0.1 pu) on one or more phase conductors for a time period between 30 cycles and 3 seconds.
- ***Interruption, Temporary :*** A type of short duration variation. The complete loss of voltage (<0.1 pu) on one or more phase conductors for a time period between 3 seconds and 1 minute.
- ***Interruption, Sustained (electric power systems) :*** Any interruption not classified as a momentary interruption.
- ***Interruption, sustained (power quality) :*** A type of long duration variation. The complete loss of voltage (<0.1 pu) on one of more phase conductors for a time greater than 1 minute.
- ***Isolated Ground :*** An insulated equipment grounding conductor run in the same conduit or raceway as the supply conductors. This conductor is insulated from the metallic

raceway and all ground points throughout its length. It originates at an isolated ground type receptacle or equipment input terminal block and terminates at the point where neutral and ground are bonded at the power source.

- ***Isolation :*** Separation of one section of a system from undesired influences of other sections.
- ***Linear Load :*** An electrical load device which, in steady state operation, presents essentially constant load impedance to the power source throughout the cycle of applied voltage.
- ***Long Duration Variation :*** A variation of the rms value of the voltage from nominal voltage for a time greater than one minute. Usually further described using a modifier indicating the magnitude of a voltage variation (*e.g.*, Undervoltage, Overvoltage, or Voltage Interruption).
- ***Low-Side Surges :*** A term coined by distribution transformer designers to describe the current surge that appears to be injected into the transformer secondary terminals upon a lightning strike to grounded conductors in the vicinity.
- ***Momentary :*** When used to quantify the duration of a short duration variation as a modifier, refers to a time range at the power frequency from 30 cycles to 3 seconds.
- ***Noise :*** Unwanted electrical signals which produce undesirable effects in the circuits of the control systems in which they occur. (For this document, "control systems" is intended to include sensitive electronic equipment in total or in part).
- ***Nominal Voltage (Vn) :*** A nominal value assigned to a circuit or system for the purpose of conveniently designating its voltage class (as 208/120, 480/277, 600).
- ***Nonlinear Load :*** Electrical load which draws current discontinuously or whose impedance varies throughout the cycle of the input ac voltage waveform.
- ***Normal Mode Voltage :*** A voltage that appears between or among active circuit conductors.
- ***Notch :*** A switching (or other) disturbance of the normal power voltage waveform, lasting less than a half-cycle; which is initially of opposite polarity than the waveform, and is thus subtracted from the normal waveform in terms of the peak value of the disturbance voltage. This includes complete loss of voltage for up to a half cycle.
- ***Oscillatory Transient :*** A sudden, non-power frequency change in the steady state condition of voltage or current that includes both positive or negative polarity value.
- ***Overvoltage :*** When used to describe a specific type of long duration variation refers to a voltage having a value of at least 10% above the nominal voltage for a period of time greater than 1 minute.
- ***Passive Filter :*** A combination of inductors, capacitors, and resistors designed to eliminate one or more harmonics. The most common variety is simply an inductor in series with a shunt capacitor, which short-circuits the major distorting harmonic component from the system.
- ***Phase Shift :*** The displacement in time of one voltage-waveform relative to other voltage waveform(s).
- ***Power Factor**, *Displacement :*** The power factor of the fundamental frequency components of the voltage and current waveforms.

- ***Power Factor (True) :*** The ratio of active power (watts) to apparent power (volt-amperes).
- ***Pulse :*** An abrupt variation of short duration of a physical quantity followed by a rapid return to the initial value.
- ***Reclosing :*** It is the common utility practice on overhead lines of closing the breaker within a short time after clearing a fault taking advantage of the fact that most faults are transient, or temporary.
- ***Recovery Time :*** Time interval needed for the output voltage or current to return to a value within the regulation specification after a step load or line change. Also may indicate the time interval required to bring a system back to its operating condition after an interruption or dropout.
- ***Recovery Voltage :*** The voltage that occurs across the terminals of a pole of a circuit interrupting device upon interruption of the current.
- ***Safety Ground :*** See : Equipment Grounding Conductor.
- ***Sag :*** A decrease to between 0.1 and 0.9 pu in rms voltage or current at the power frequency for durations of 0.5 cycles to one minute.
- ***Shield :*** As normally applied to instrumentation cables, refers to a conductive sheath (usually metallic) applied, over the insulation of a conductor or conductors, for the purpose of providing means to reduce coupling between the conductors so shielded and other conductors which may be susceptible to, or which may be generating unwanted electrostatic or electromagnetic fields (noise).
- ***Shielding :*** Shielding is the use of a conducting and/or ferromagnetic barrier between a potentially disturbing noise source and sensitive circuitry. Shields are used to protect cables (data and power) and electronic circuits. They may be in the form of metal barriers, enclosures, or wrappings around source circuits and receiving circuits.
- ***Shielding (of utility lines) :*** The construction of a grounded conductor or tower over the lines to intercept lightning strokes in an attempt to keep the lightning currents out of the power system.
- ***Short Duration Variation :*** A variation of the rms value of the voltage from nominal voltage for a time greater than one-half cycle of the power frequency but less than or equal to one minute. Usually further described using a modifier indicating the magnitude of a voltage variation (*e.g.*, Sag, Swell, or Interruption) and possibly a modifier indicating the duration of the variation (*e.g.*, Instantaneous, Momentary or Temporary).
- ***Signal Reference Grid (or Plane) :*** A system of conductive paths among interconnected equipment, which reduces noise-induced voltages to levels which minimize improper operation. Common configurations include grids and planes.
- ***Sustained :*** When used to quantify the duration of a voltage interruption, refers to the time frame associated with a long duration variation (*i.e.*, greater than one minute).
- ***Swell :*** A temporary increase in the rms value of the voltage of more than 10% the nominal voltage, at the power frequency, for durations from 0.5 cycle to one minute.
- ***Synchronous Closing :*** Generally used in reference to closing all three poles of a capacitor switch in synchronism with the power system to minimize transients.
- ***Temporary :*** When used to quantify the duration of a short duration variation as a modifier, refers to a time range from 3 seconds to 1 minute.

- ***Total Demand Distortion (TDD)*** **:** The ratio of the root-mean- square of the harmonic current to the root-mean-square value of the rated or maximum demand fundamental current, expressed as a percent.
- ***Total Disturbance Level*** **:** The level of a given electromagnetic disturbance caused by the superposition of the emission of all pieces of equipment in a given system.
- ***Total Harmonic Distortion (THD)*** **:** The ratio of the root-mean- square of the harmonic content to the root-mean-square value of the fundamental quantity, expressed as a percent of the fundamental.
- ***Transient*** **:** Pertaining to or designating a phenomenon or a quantity which varies between two consecutive steady states during a time interval that is short compared to the time scale of interest. A transient can be a unidirectional impulse of either polarity or a damped oscillatory wave with the first peak occurring in either polarity.
- ***Triplen Harmonics*** **:** A term frequently used to refer to the odd multiples of the third harmonic, which deserve special attention because of their natural tendency to be zero sequence.
- ***Undervoltage*** **:** When used to describe a specific type of long duration variation, refers to a measured voltage having a value at least 10% below the nominal voltage for a period of time greater than one minute.
- ***Voltage Change*** **:** A variation of the rms or peak value of a voltage between two consecutive levels sustained for definite but unspecified durations.
- ***Voltage Dip*** **:** See Sag.
- ***Voltage Distortion*** **:** Distortion of the ac line voltage. See Distortion.
- ***Voltage Fluctuation*** **:** A series of voltage changes or a cyclical variation of the voltage envelope.
- ***Voltage Imbalance (Unbalance)*** **:** It is a condition in which the three phase voltages differ in amplitude or are displaced from their normal 120 degree phase relationship or both. It is frequently expressed as the ratio of the negative sequence or zero sequence voltage to the positive sequence voltage, in percent.
- ***Voltage Interruption*** **:** Disappearance of the supply voltage on one or more phases. Usually qualified by an additional term indicating the duration of the interruption (*e.g.*, Momentary, Temporary, or Sustained.)
- ***Voltage Regulation*** **:** The degree of control or stability of the rms voltage at the load. Often specified in relation to other parameters, such as input-voltage changes, load changes, or temperature changes.
- ***Voltage Magnification*** **:** The magnification of capacitor switching oscillatory transient voltage on the primary side by capacitors on the secondary side of a transformer.
- ***Waveform Distortion*** **:** A steady-state deviation from an ideal sine wave of power frequency principally characterized by the spectral content of the deviation.

❑❑❑

INDEX

N

O

P

R

S

T

U

V

Z

□□□